AF357842

Innovative Material Design and Nondestructive Testing Applications for Infrastructure Materials

Innovative Material Design and Nondestructive Testing Applications for Infrastructure Materials

Guest Editors

Honglei Chang
Li Ai
Feng Guo
Xinxiang Zhang

Basel • Beijing • Wuhan • Barcelona • Belgrade • Novi Sad • Cluj • Manchester

Guest Editors

Honglei Chang
School of Qilu Transportation
Shandong University
Jinan
China

Li Ai
Department of Mechanical
Engineering
University of South Carolina
Columbia
United States

Feng Guo
School of Qilu Transportation
Shandong University
Jinan
China

Xinxiang Zhang
Department of Electrical and
Computer Engineering
Southern Methodist
University
Dallas
United States

Editorial Office
MDPI AG
Grosspeteranlage 5
4052 Basel, Switzerland

This is a reprint of the Special Issue, published open access by the journal *Materials* (ISSN 1996-1944), freely accessible at: www.mdpi.com/journal/materials/special_issues/6UFRF89J27.

For citation purposes, cite each article independently as indicated on the article page online and using the guide below:

Lastname, A.A.; Lastname, B.B. Article Title. *Journal Name* **Year**, *Volume Number*, Page Range.

ISBN 978-3-7258-3334-4 (Hbk)
ISBN 978-3-7258-3333-7 (PDF)
https://doi.org/10.3390/books978-3-7258-3333-7

Contents

Editorial

Special Issue: Innovative Material Design and Nondestructive Testing Applications for Infrastructure Materials

Honglei Chang * and **Feng Guo**

School of Qilu Transportation, Shandong University, Jinan 250003, China; fengg@sdu.edu.cn
* Correspondence: hlchang@sdu.edu.cn

check for updates

Received: 6 January 2025
Revised: 22 January 2025
Accepted: 27 January 2025
Published: 29 January 2025

Citation: Chang, H.; Guo, F. Special Issue: Innovative Material Design and Nondestructive Testing Applications for Infrastructure Materials. *Materials* **2025**, *18*, 611.
https://doi.org/10.3390/ma18030611

Construction materials play a vital role in the design, construction, and maintenance of transportation infrastructure, significantly impacting its safety, durability, and stability. The rapid development of global urbanization and the increasing demand for transportation intensified challenges, including aging materials, environmental degradation, and the impacts of climate change. To address these issues, the development of innovative materials and advanced nondestructive testing techniques is essential. Researchers are actively developing new construction materials, enhancing the performance of existing ones, and advancing cutting-edge nondestructive testing methods to meet the evolving requirements of modern, increasingly complex transportation infrastructure.

This Special Issue, titled "Innovative Material Design and Nondestructive Testing Applications for Infrastructure Materials," showcases cutting-edge research and technological advancements in the field of infrastructure materials. Featuring ten high-quality research articles developed over nearly a year, this collection covers a diverse range of topics in innovative material design and nondestructive testing techniques. Contributions from more than 40 authors representing various organizations provide comprehensive insights and perspectives on this rapidly evolving and critical area of study.

Dong et al. [1] investigated the impact of coarse particles, such as sand and walnut shells, on filter cake formation and air tightness in high-permeability formations. Their findings revealed that walnut shells were more effective than sand in improving the airtightness of filter cakes, and they recommended incorporating walnut shells at a rate of 30–40 g/L for formations with permeability coefficients greater than 1.0×10^{-3} m/s. Nering and Nering [2] assessed vibration isolation materials for reducing urban noise and vibrations, employing a single-degree-of-freedom system and image-processing techniques. Their study demonstrated that the image-processing method reliably predicted the materials' dynamic stiffness and damping properties, revealing strong correlations between indentation and dynamic stiffness, as well as rebound and damping. Haruna et al. [3] studied the bonding behavior of NSC-UHPFRC composites under impact loading. Their findings revealed that surface treatment methods significantly influenced bond strength, and the XGBoost model accurately predicted impact strength. Liu et al. [4] explored the effects of freeze–thaw cycling and cyclic loading on the damage evolution of sandstone. The results showed that freeze–thaw cycling increased sandstone's total porosity and microporosity in a linear fashion. Using the loading–unloading response ratio and the strain equivalence principle, a damage model was developed for fractured rock under freeze–thaw–fatigue coupling. Peng et al. [5] examined the bond degradation between polyurea coatings and concrete substrates using accelerated aging experiments. They found that bond strength declined over time and developed an aging model to predict the coating's service life.

Fu et al. [6] analyzed sandstone damage evolution under high strain rates using a split Hopkinson pressure bar. They observed that internal stress inhomogeneity remained

https://doi.org/10.3390/ma18030611

within 5% across strain rates of 931–2250 s^{-1}, highlighting the influence of deformation stages and strain rates on material behavior. Behnia and Lukaszewski [7] proposed a novel method combining acoustic emission, machine learning, and digital image techniques to analyze spiral crack patterns in soft binders. The study found strong correlations between helical crack energy and properties such as fracture energy and embrittlement temperature. He et al. [8] investigated the axial compressive properties of recycled aggregate concrete reinforced with ultra-high toughness cementitious composites. Steel fibers significantly enhanced the mechanical properties of the structures, with the study recommending their use for applications requiring higher mechanical strength. Li et al. [9] developed a pavement icing detection system based on a piezoelectric sensor, integrated with a BP neural network for early warning. The system achieved a prediction accuracy exceeding 90%, offering an effective approach for winter road safety monitoring. Ji et al. [10] analyzed the permanent deformation of asphalt pavement-bearing layers and proposed a control criterion based on dynamic modulus and stability. This approach aims to improve the deformation resistance of bearing layers, thereby extending the pavement's service life.

As a multidisciplinary and expansive topic, "Innovative Material Design and Nondestructive Testing Applications for Infrastructure Materials" cannot be comprehensively explored within the scope of a single Special Issue. Nevertheless, the papers included in this issue provide valuable insights into some of the critical challenges in the field. They highlight the diverse applications of innovative material design and nondestructive testing techniques in infrastructure. It is our hope that these findings will serve as a foundation for further research and inspire continued advancements, contributing to the progress of infrastructure materials science and engineering.

Author Contributions: Conceptualization, H.C.; investigation, H.C. and F.G.; resources, H.C. and F.G.; writing—original draft preparation, H.C.; writing—review and editing, F.G.; visualization, H.C. and F.G.; supervision, H.C.; project administration, F.G. All authors have read and agreed to the published version of the manuscript.

Conflicts of Interest: The authors declare no conflicts of interest.

References

1. Dong, Q.; Liu, T.; Wang, Y.; Liu, S.; Wen, L. Study on the Influence of Walnut Shell Coarse Particles on the Slurry Permeation and the Air Tightness of Filter Cake. *Materials* **2024**, *17*, 5186. [CrossRef] [PubMed]
2. Nering, K.; Nering, K. Alternative Method for Determination of Vibroacoustic Material Parameters for Building Applications. *Materials* **2024**, *17*, 3042. [CrossRef] [PubMed]
3. Haruna, S.I.; Ibrahim, Y.E.; Hassan, I.H.; Al-shawafi, A.; Zhu, H. Bond Strength Assessment of Normal Strength Concrete–Ultra-High-Performance Fiber Reinforced Concrete Using Repeated Drop-Weight Impact Test: Experimental and Machine Learning Technique. *Materials* **2024**, *17*, 3032. [CrossRef] [PubMed]
4. Liu, T.; Cai, W.; Sheng, Y.; Huang, J. Experimental Study on the Microfabrication and Mechanical Properties of Freeze–Thaw Fractured Sandstone under Cyclic Loading and Unloading Effects. *Materials* **2024**, *17*, 2451. [CrossRef] [PubMed]
5. Peng, C.; Ren, J.; Wang, Y. Degradation Behavior and Lifetime Prediction of Polyurea Anti-Seepage Coating for Concrete Lining in Water Conveyance Tunnels. *Materials* **2024**, *17*, 1782. [CrossRef]
6. Fu, Y.; Chen, S.; Zhao, P.; Ye, X. The Mechanism of Deformation Compatibility of TA2/Q345 Laminated Metal in Dynamic Testing with Split-Hopkinson Pressure Bar. *Materials* **2023**, *16*, 7659. [CrossRef] [PubMed]
7. Behnia, B.; Lukaszewski, M. Novel Approach in Fracture Characterization of Soft Adhesive Materials Using Spiral Cracking Patterns. *Materials* **2023**, *16*, 7412. [CrossRef] [PubMed]
8. He, L.; Peng, S.; Jia, Y.-S.; Yao, Y.-K.; Huang, X.-W. Testing and Analysis of Ultra-High Toughness Cementitious Composite-Confined Recycled Aggregate Concrete under Axial Compression Loading. *Materials* **2023**, *16*, 6573. [CrossRef]

9.   Li, J.; Ma, H.; Shi, W.; Tan, Y.; Xu, H.; Zheng, B.; Liu, J. Nondestructive Detection and Early Warning of Pavement Surface Icing Based on Meteorological Information. *Materials* **2023**, *16*, 6539. [CrossRef] [PubMed]

10.   Ji, W.; Meng, Y.; Shang, Y.; Zhou, X.; Xu, H. Investigation of the Relationship between Permanent Deformation and Dynamic Modulus Performance for Bearing-Layer Asphalt Mixture. *Materials* **2023**, *16*, 6409. [CrossRef] [PubMed]

 *materials*

Article

Study on the Influence of Walnut Shell Coarse Particles on the Slurry Permeation and the Air Tightness of Filter Cake

Qi Dong [1], Tao Liu [2,*], Yuan Wang [1], Sijin Liu [3] and Letian Wen [4]

1. College of Water Conservancy and Hydropower Engineering, Hohai University, Nanjing 210024, China; dong0626qi@hhu.edu.cn (Q.D.)
2. College of Civil Engineering and Transportation, Hohai University, Nanjing 210024, China
3. China Railway 14th Bureau Group Co., Ltd., Jinan 250000, China
4. China South-to-North Water Diversion Middle Route Co., Ltd., Beijing 100036, China
* Correspondence: 211604010111@hhu.edu.cn; Tel.: +86-13105273696

Abstract: Slurry shields rely on the formation of a compact filter cake to maintain excavation face stability and ensure construction safety. In strata with high permeability, significant slurry loss occurs, making filter cake formation and air tightness maintenance challenging. In this study, light organic walnut shell was selected as an additive coarse particle material for slurry. Slurries incorporating two types of coarse particles, sand and walnut shell, were prepared, and tests on slurry permeation and air tightness of the filter cake were conducted in three different strata. The results indicate that the addition of coarse particles effectively improves filter cake formation and air tightness in high-permeability strata. It is essential to use graded particles in highly permeable strata, with controlled maximum and minimum particle sizes. As the content of coarse particles increases, the air tightness of the filter cake initially increases and then decreases. Notably, the air tightness of filter cakes containing walnut shell is superior to those containing sand. Replacing sand with walnut shell as a slurry plugging material enhances filter cake quality in high-permeability strata. For highly permeable strata with a permeability coefficient greater than 1.0×10^{-3} m/s, an addition of 30 g/L to 40 g/L is recommended.

Keywords: slurry shield; coarse particle; walnut shells; slurry infiltration; filter cake; air tightness

Citation: Dong, Q.; Liu, T.; Wang, Y.; Liu, S.; Wen, L. Study on the Influence of Walnut Shell Coarse Particles on the Slurry Permeation and the Air Tightness of Filter Cake. *Materials* **2024**, *17*, 5186. https://doi.org/10.3390/ma17215186

Academic Editors: Eddie Koenders, Maria Stefanidou and Marco Corradi

Received: 9 June 2024
Revised: 29 August 2024
Accepted: 3 October 2024
Published: 24 October 2024

1. Introduction

The slurry shield method is widely applied in the construction of underwater tunnels [1,2]. The filter cake formed by the infiltration of slurry into the excavation face acts as the first barrier to maintain the stability of the excavation face and ensure construction safety, making it a focal point in related research [3,4]. The pressurized slurry in the excavation chamber infiltrates forward, with the solid phase in the slurry gradually filling the surface pores of the strata to form a slightly permeable filter cake. The slurry pressure acts on the excavation face through the filter cake in the form of surface force, balancing the external soil and water pressure of the strata [5]. Talmon et al. [6] proposed that the condition for filter cake formation is that the particle deposition rate is greater than the infiltration rate of the slurry into the strata. Kou et al. [7] identified three infiltration modes and the deposition patterns of slurry filter cake formation through extensive experiments, each corresponding to different types of infiltration curves. On this basis, many scholars have studied the impact of slurry properties [8,9] on the formation effectiveness of filter cakes and have attempted to enhance the quality of the filter cake by using various slurry additives [10–13].

In low-permeability strata, the expansion and cementation of small particles in pure bentonite slurry are sufficient to form a filter cake. However, in highly permeable strata (such as sand and gravel layers), the large pores of the strata cause significant loss of pure bentonite slurry, making filter cake formation difficult. Therefore, a certain amount of coarse particles must be added to the slurry to block the pores of the strata. The classical filtration model theoretically explains the deposition and growth of the filter cake [14]. Tien et al. have improved the conventional cake filtration model based on multiphase flow equations [15], and Herzig et al. [16] proposed the mechanism for particle clogging and retention based on this model. In practice, the formation of a high-quality filter cake requires good compatibility between the slurry particles and the strata. Traditional theories generally classify filter cake types based on the ratio of particle diameter in the slurry to the pore diameter of the strata. Common matching criteria include the '1/3, 1/7 rule' and the '1/3, 1/14 rule' [17]. Min et al. [18] proposed a formation compatibility criterion based on the average pore diameter of the strata and the d_{85} of the slurry. In the field of shield tunneling construction, the most commonly used coarse particles are quartz sand [19–21]. Lin et al. suggested that the slurry filtration process starts with the large particles, and sand-added slurry can significantly accelerate the formation of filter cakes in coarse strata [22]. Ma et al. proposed that the solid particle size of sands has a significant effect on the filter cake growth process [23]. However, due to its high specific gravity, quartz sand tends to settle at the bottom of the excavation chamber, leading to an uneven filter cake, which fails to meet the quality requirements for shield machine tunneling and chamber opening. In recent years, organic coarse particle materials such as natural nutshells [24,25], plant fibers [26,27], and eggshells [28,29], which are used as lost circulation materials in drilling slurry, have attracted attention and have been increasingly used in shield tunneling construction due to their better compatibility with the slurry [30].

Additionally, as the shield tunneling distance increases, the tunnel boring machine (TBM) often requires chamber inspections due to tool wear or encountering boulders [31]. To maintain the stability of the excavation face during pressurized chamber openings, the filter cake must possess a certain degree of air tightness, which imposes higher quality requirements on the formation of filter cakes [32,33]. The failure of the filter cake under certain air pressure conditions is mainly due to the ingress of air, which occupies the pores of the filter cake and causes a sharp decrease in its internal saturation [34–37]. Additionally, with increasing air pressure and its duration, the pore structure of the filter cake changes [38]. The commonly used evaluation indicators for the air tightness of the filter cake are the air tightness pressure value and air tightness time [39]. The pore structure, thickness of the filter cake, permeability coefficient of the strata [40], and the content of coarse particles [41] all affect the air tightness performance of the filter cake.

To address the issue of poor filter cake quality caused by significant settling of sand particles in traditional sand-containing slurry in highly permeable strata, lightweight organic walnut shells were selected as slurry plugging additives due to their good plugging effect and durability [42]. Comparative tests on slurry infiltration and the air tightness of the filter cake were conducted using both sand-containing and walnut shell slurries in three sand layers of different permeabilities. This study analyzed the influence and mechanisms of coarse particle type, particle size, particle content, and strata permeability coefficient on the formation and air tightness of filter cakes. The findings aim to provide guidance for the application of walnut shells in shield tunneling and pressurized maintenance operations in highly permeable strata.

2. Experimental Materials and Methods

2.1. Experimental Materials

2.1.1. Strata Materials

To study the slurry infiltration and filter cake formation in highly permeable sand layers with different pore structures, three types of strata with varying permeability were prepared using well-graded natural river sand. These strata, ordered by increasing permeability coefficient, are S1 (grain size 0.5–1 mm), S2 (grain size 2–5 mm), and S3 (grain size 3–5 mm). The basic parameters of each strata are listed in Table 1. The average pore diameter of the strata is calculated based on the characteristic pore value proposed by Min et al. [18]. The permeability coefficients of the strata were measured using the constant head permeability test.

Table 1. Strata parameters in experiments.

Strata Number	Dry Density/(g/cm^3)	Porosity	Average Pore Size/(mm)	Permeability Coefficient/(m/s)
S1	1.441	0.340	0.145	5.04×10^{-3}
S2	1.469	0.336	0.464	5.07×10^{-2}
S3	1.510	0.330	1.002	2.30×10^{-1}

2.1.2. Slurry Materials

The materials used for slurry preparation in the experiments mainly included purified water, sodium-based bentonite, carboxymethyl cellulose (CMC), walnut shell particles, and sand particles. The base slurry was prepared at a water-to-bentonite ratio of 1:8, with a certain mass fraction of CMC added as a thickening agent. Each group of slurry was formulated by adding different types and amounts of coarse particles to the base slurry. The sodium-based bentonite was sieved using a 200-mesh (0.075 mm) standard sieve (Huafeng Hardware Instruments Co., Ltd., Shaoxing, China) to remove impurities. The particle size of the coarse particles was determined based on the d_{85} criterion proposed by Min et al. [18] and the standard sieve apertures used in the field. Coarse particles were added as particle groups rather than single particle sizes. The experimental groups and basic properties of the slurries are shown in Table 2.

Table 2. Basic parameters of slurry properties.

Slurry Number	Coarse Particle	Coarse Particle Size/mm	Added Amount of Coarse Particle/(g/L)	Added Amount of CMC/‰	24 h Funnel Viscosity/s	Specific Gravity /(g/cm^3)
S1-0	none	—	0		35	1.075
S1-H-1			10		55	1.082
S1-H-2			20		61	1.086
S1-H-3	walnut shell	0.075–0.15	30		65	1.090
S1-H-4			40	1	68	1.092
S1-S-1			10		49	1.082
S1-S-2			20		52	1.093
S1-S-3	sand	0.075–0.15	30		57	1.100
S1-S-4			40		51	1.106
S2-0	none	—	0		42	1.086
S2-H-1			10		62	1.090
S2-H-2			20		78	1.093
S2-H-3	walnut shell	0.25–0.5	30		82	1.095
S2-H-4			40	2	84	1.097
S2-S-1			10		50	1.091
S2-S-2			20		51	1.097
S2-S-3	sand	0.25–0.5	30		52	1.100
S2-S-4			40		48	1.111
S3-0	none	—	0		60	1.088
S3-H-1			10		78	1.094
S3-H-2		0.25–1	20		77	1.096
S3-H-3			30		85	1.097
S3-H-4	walnut shell		40		89	1.100
S3-H-5		0.25–0.5	30		85	1.095
S3-H-6		0.5–1	30	3	85	1.097
S3-S-1			10		63	1.097
S3-S-2		0.25–1	20		61	1.102
S3-S-3	sand		30		63	1.103
S3-S-4			40		65	1.106
S3-S-5		0.25–0.5	30		63	1.103
S3-S-6		0.5–1	30		63	1.103

2.2. Test Apparatus

The experiment utilized a self-made test system of slurry permeation and filter cake air tightness, comprising three main components: the pressurization system, the permeation column, and the collection system of permeation flow rate. The pressurization system consisted of a pressurization device (air compressor, Aotus Industry and Trade Co., Ltd., Taizhou, China) and a pressure regulation device (pressure regulator valve and bidirectional pressure gauge, Bowei Pressure Regulator Factory, Taizhou, China). The permeation column was made of organic glass with an inner diameter of 10 cm and a height of 70 cm, sealed at the top by a flange plate and connected to the flow rate collection device at the bottom via an open end. The permeation flow rate collection system included a high-precision balance, a filtrate collector, and a data acquisition software program (Anheng Flow Data Acquisition Software, and the version number is ACS-30), as shown in Figure 1.

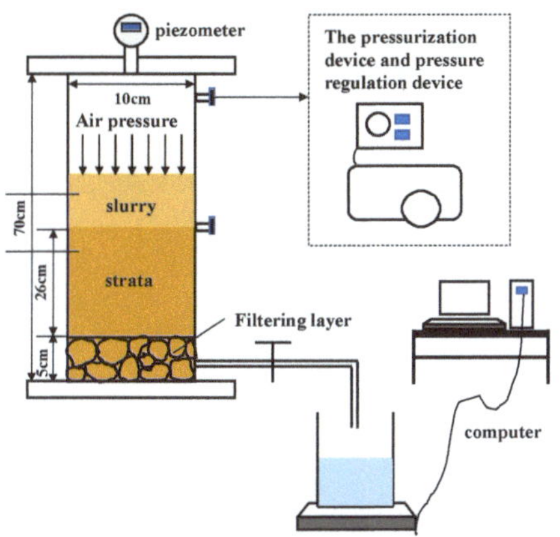

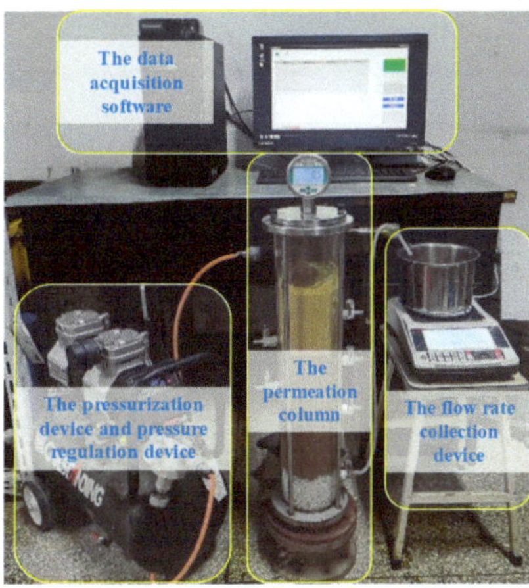

Figure 1. Test system of slurry permeation and filter cake air tightness.

2.3. Experimental Methods

Firstly, the strata filling process was initiated, with a bottom filter layer comprising white gravel with a particle size range of 6–8 mm and a height of 5 cm. The experimental strata consisted of well-graded natural river sand with a height of 26 cm. The dry density of the strata was strictly controlled to be the same for each test group. After filling the strata, the saturation process was conducted from bottom to top for 24 h. Once saturation was achieved, 1000 mL of prepared slurry was slowly injected. The slurry permeation test commenced after a 5 min settling period.

The slurry permeation test was conducted using a staged pressurization method, with pressures of 50 kPa, 100 kPa, 150 kPa, and 200 kPa applied successively, each maintained for 3 min. Throughout the experiment, the water outlet valve at the bottom of the permeation column remained open, and the data acquisition system automatically recorded the flow rate at 1 s intervals. After the test, the permeation column was slowly depressurized, and remaining slurry was drained by opening the valves on the side of the permeation column. The filter cake morphology was observed. The thickness and penetration distance of the filter cake were measured.

For the slurry groups that stable filter cakes formed during the permeation test, air tightness tests of the filter cakes were conducted. Each slurry group underwent two air tightness tests to determine the air tightness pressure value and air tightness time. After the formation of the filter cake, any remaining slurry in the permeation column was drained to allow direct air pressure on the filter cake surface. Air tightness pressure values were determined using staged pressurization. The pressure was incrementally increased by

30 kPa until filter cake failure occurred, with each pressure level maintained for 1.0 min. Air tightness time was measured under a constant pressure of 30 kPa until failure of the filter cake. Flow rate monitoring was conducted throughout the test, and the filter cake was removed after completion and the failure morphology was observed.

3. Results

3.1. Impact of Coarse Particle Materials on the Formation Characteristics of Filter Cake

3.1.1. Influence of Coarse Particle Size on Filter Cake Formation

The types of low-permeability zones formed by the permeation of various experimental slurries are shown in Table 3. Two types of low-permeability zones were observed, which were the combination type of filter cake and permeation band, as shown in Figure 2, and penetrating-type permeation bands.

Table 3. Summary of slurry permeation phenomena.

Strata Number	Slurry Type	Type of Low-Permeability Zones
S1	no coarse particles containing 0.075–0.15 mm coarse particles	penetrating-type permeation bands filter cake and permeation band
S2	no coarse particles containing 0.25–0.5 mm coarse particles	penetrating-type permeation bands filter cake and permeation band
S3	no coarse particles containing 0.5–1.0 mm coarse particles	penetrating-type permeation bands penetrating-type permeation bands

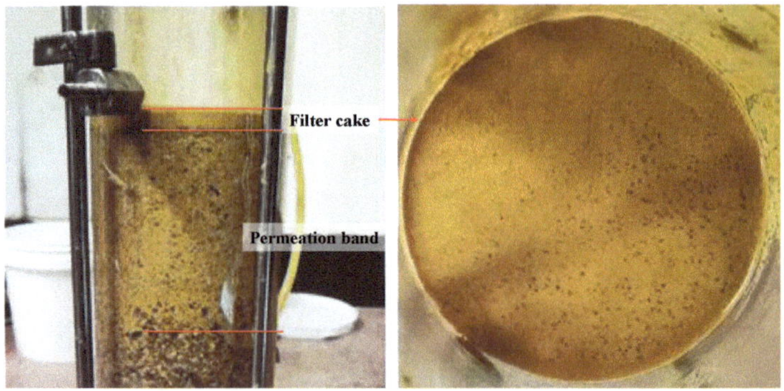

Figure 2. The combination type of permeation band and filter cake.

In all three strata, slurries without coarse particles failed to form a filter cake. According to the d_{85} matching criterion, the required particle sizes for filter cake formation in the S1, S2, and S3 strata should be 0.075–0.15 mm, 0.25–0.5 mm, and 0.5–1 mm, respectively. In the S1 and S2 strata, except for the slurry containing 10 g/L of 0.25–0.5 mm sand particles failing to form the filter cake in the S2 strata, all other slurries formed low-permeability zones characterized by the combination type of filter cake and permeation band. However, in the S3 strata, none of the slurries were able to form a filter cake. Therefore, while the particle sizes determined by the d_{85} criterion meet the requirements for filter cake formation in the relatively low-permeability S1 and S2 strata, they fail to meet the requirements for filter cake formation in the relatively high-permeability S3 strata.

To address this issue, additional permeation tests were conducted in the S3 strata using slurries supplemented with coarse particles in the particle size ranges of 0.25–0.5 mm, 0.5–1.0 mm, and 0.25–1.0 mm. The results revealed that slurries containing 0.25–0.5 mm coarse particles only formed permeation bands, with particles and slurry infiltrating the strata through its pores, as depicted in Figure 3a. Similarly, slurries containing 0.5–1.0 mm coarse particles also only formed permeation bands, with most coarse particles remaining on the surface of the strata, while fine particles in the slurry permeated through the pores formed by the accumulation of coarse particles, as illustrated in Figure 3b. However, slurries containing 0.25–1.0 mm coarse particles were able to generate a stable filter cake and permeation band combination, as shown in Figure 3c.

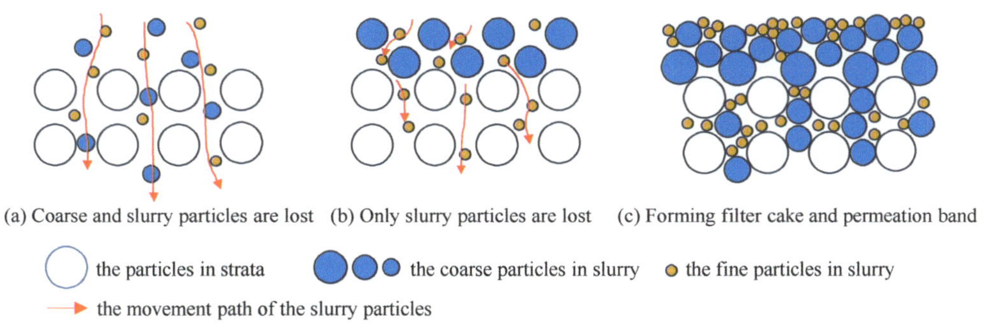

Figure 3. The bridging effect of coarse particles with different gradations in slurry.

3.1.2. Analysis of Permeation Flow in Filter Cake Formation

Under a certain pressure, the slurry infiltrates into the strata, with the liquid phase permeating into the strata while the solid phase accumulates on the surface to form a filter cake. The formation of the filter cake gradually reduces the permeability of the strata, hindering further infiltration of the liquid phase. Figure 4 depicts the variation in permeation flow during the formation of the filter cake in the slurry containing two types of coarse particles (using the S3 strata as an example).

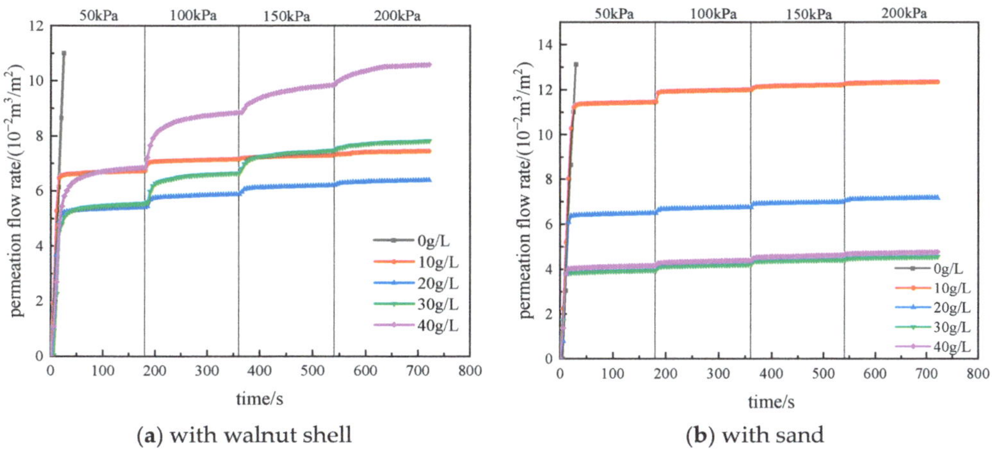

Figure 4. Infiltration flow rate curves of slurry in S3 stratums.

After the application of pressure, the slurry permeation flow initially increases rapidly and then gradually stabilizes, indicating the formation of the filter cake. The pressure acts on the strata through the filter cake in the form of effective stress. In the experiment, each increase in pressure level results in a sudden change in flow rate, mainly due to filter cake compression and local breakthroughs, which then stabilizes as the filter cake is compacted or repaired. During the incremental pressurization process, the maximum flow rate is observed at the first pressure level, and subsequent pressure increments result in smaller flow rate changes, indicating the gradual densification of the filter cake. The statistical analysis of the permeation flow rates for each group of slurry is shown in Figure 5. Except for the slurry with 10 g/L of sand in the S2 strata, which did not form a filter cake, all other slurry groups formed stable filter cakes and permeation bands.

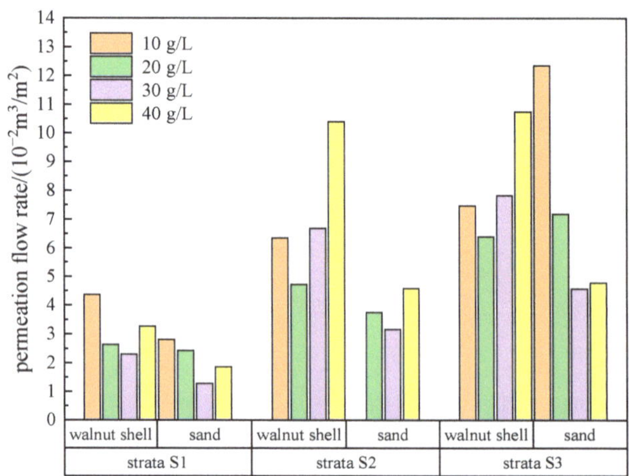

Figure 5. Infiltration flow volume of forming filter cake under varied permeability strata and coarse particle content.

When slurry with the same type of coarse particles is permeating in strata with a fixed permeability coefficient, the permeation flow rate initially decreases and then increases with the increase in the amount of coarse particle addition, reaching a minimum value at a coarse particle content of 20 g/L or 30 g/L. Since this experiment is for vertical permeation, the permeation flow rate of slurry containing sand particles is slightly lower than that of slurry containing walnut shell particles. This is because sand particles have a higher density and can reach the strata surface more quickly under the action of gravity to block it, resulting in a shorter filter cake formation time and lower permeation flow rate. However, in actual construction, the permeation is lateral, and the density difference between the base slurry and sand particles can cause particle sedimentation, resulting in uneven filter cake quality. Additionally, as the permeability coefficient of the strata increases, the permeation flow rate of slurry containing both types of particles also increases.

3.1.3. Permeability Analysis of Filter Cakes

The permeability coefficient of the filter cake is one of the key indicators to measure the quality of the filter cake. The process of slurry infiltration can be simplified into a one-dimensional seepage model. According to Equation (1), the permeability coefficient of the filter cake can be expressed as

$$K = \frac{qL}{Ah} \tag{1}$$

where K represents the permeability coefficient of the filter cake, measured in m/s; q denotes the flux rate per unit time, measured in m^3/s; A stands for the cross-sectional area of the sample, measured in m^2, which corresponds to the cross-sectional area of the permeation column in the experiment, taken as 7.85×10^{-3} m^2; h represents the water level difference at both ends of the sample, measured in m, which is taken as the final pressure difference during the slurry infiltration to form the filter cake in the experiment, specifically 20 m; and L stands for the length of the permeation path, which in the experiment is taken as the total thickness of the filter cake, measured in m.

The permeability coefficients of the filter cakes in each group are shown in Table 4. The permeability coefficients of the filter cakes formed by slurry infiltration in each group have reached the order of magnitude of 10^{-7} m/s or even smaller, effectively blocking the pores of the strata and forming low-permeability zones, meeting the requirements of shield tunneling construction.

Table 4. Permeability coefficients of the filter cakes.

Slurry Number	Permeability Coefficient of the Filter /(10^{-7} m/s)	Slurry Number	Permeability Coefficient of the Filter/(10^{-7} m/s)	Slurry Number	Permeability Coefficient of the Filter/(10^{-7} m/s)
S1-H-1	0.92	S2-H-1	1.83	S3-H-1	1.49
S1-H-2	1.60	S2-H-2	2.04	S3-H-2	2.29
S1-H-3	2.29	S2-H-3	4.71	S3-H-3	5.25
S1-H-4	3.37	S2-H-4	7.12	S3-H-4	7.54
S1-S-1	0.79	S2-S-1	—	S3-S-1	1.07
S1-S-2	1.25	S2-S-2	1.12	S3-S-2	1.68
S1-S-3	1.56	S2-S-3	1.07	S3-S-3	2.10
S1-S-4	2.58	S2-S-4	1.91	S3-S-4	2.98

When slurry with the same type of coarse particles permeates the same strata, the higher the amount of particle addition, the greater the permeability coefficient of the formed filter cake. This is because with more particles, the resulting filter cake becomes thicker and more porous, enhancing its compressibility. Consequently, during permeation, the increase in permeation flow rate due to the compression of the filter cake leads to longer permeation distances, resulting in a higher permeability coefficient. Therefore, during shield tunneling operations involving pressurized chamber breakthroughs, higher permeation pressure and longer permeation time are advised to allow for sufficient compression of the filter cake, thus obtaining a higher-quality filter cake.

3.2. Impact of Coarse Particle Materials on the Air Tightness Characteristics of Filter Cakes

3.2.1. Air Tightness Pressure Value of Filter Cake

The air tightness pressure value denotes the maximum air pressure that a filter cake can withstand while preserving its air tightness. This parameter is crucial for assessing the filter cake's capacity to endure elevated air pressures. During pressurized maintenance operations, the set air tightness differential—defined as the difference between the working pressure and external water pressure—must not exceed the filter cake's air tightness pressure value.

In the experiment, excess slurry was drained, and incremental air pressure was applied until the filter cake failed. The failure of the filter cake under high air pressure can occur either at the moment of pressure increase at a certain pressure level or during constant pressure, manifesting as sudden instantaneous failure. The failure morphology typically appears as a circular hole breakthrough, as shown in Figure 6.

The permeation flow rate curve of the filter cake, from incremental pressure application to failure, is shown in Figure 7 (using S1 strata as an example). During each pressure level maintenance, the displacement of pore water due to air pressure causes a linear increase in the permeation flow rate of the filter cake. At the moment when the next level of air pressure load is applied, there is a slight sudden increase in permeation flow rate, primarily due to filter cake compression and drainage. After stabilization, the permeation flow rate continues to increase linearly. Once the critical point is reached, the filter cake's ability to withstand high air pressure reaches its limit, resulting in air tightness failure.

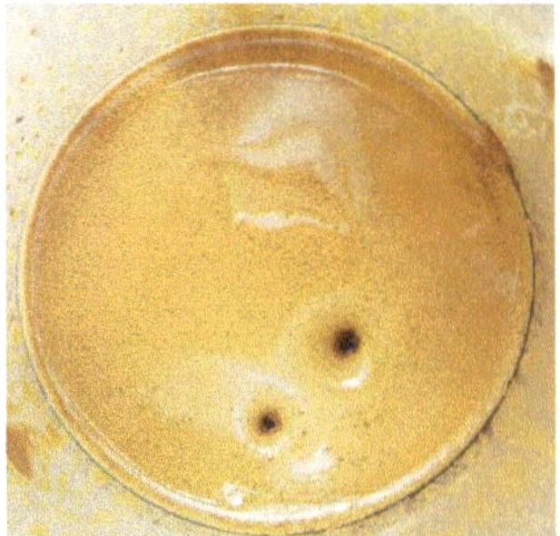

Figure 6. The failure morphology of filter cake in high air pressure.

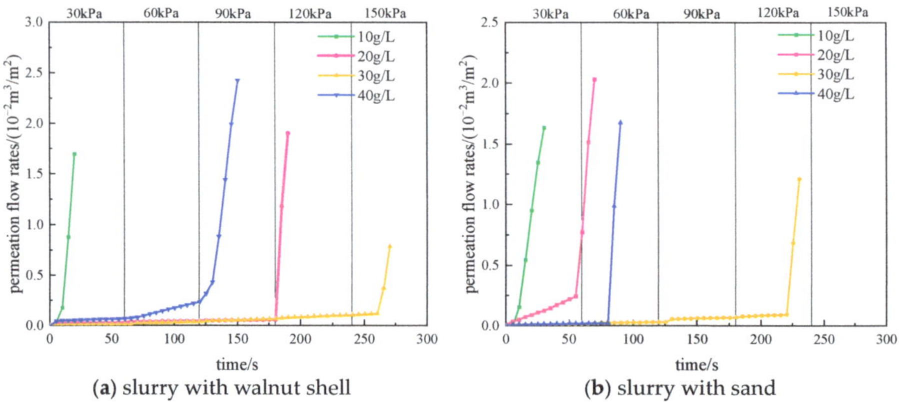

(**a**) slurry with walnut shell   (**b**) slurry with sand

Figure 7. The permeation flow rate curve of the filter cake during air tightness process.

The pressure level just before the failure point is considered as the air tightness pressure value of the filter cake. The air tightness pressure values of the filter cake for each group of slurries are statistically analyzed, as shown in Figure 8. As the strata permeability increases, it becomes more difficult to achieve air tightness, resulting in fewer successful air tightness outcomes and smaller air tightness pressure values for the filter cake groups. Additionally, with the enlargement of strata pores, filter cakes with higher levels of coarse particle addition gradually demonstrate advantages.

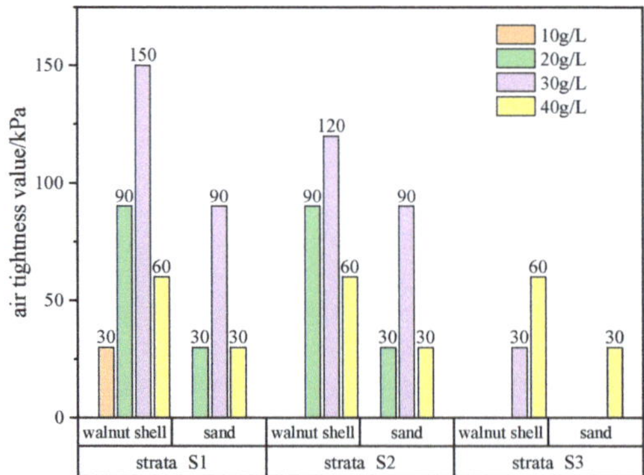

Figure 8. The air tightness pressure of the filter cake under varied permeability strata and coarse particle content.

Within the same stratum, as the particle addition increases, the air tightness pressure of the filter cake initially rises and then declines. Filter cakes containing two types of particles in the S1 and S2 strata both reach their maximum air tightness pressure value at a content of 30 g/L. In the S3 stratum, only three groups of slurries successfully achieve air tightness. Filter cakes containing walnut shells reach their maximum air tightness pressure value at a content of 40 g/L, while sand-containing slurries only achieve successful air tightness at a content of 40 g/L. Comparing the two types of added particles, under the same particle addition level, the air tightness pressure value of filter cakes containing walnut shells is larger than that of filter cakes containing sand, indicating that filter cakes generated from walnut shell-containing slurries have better resistance to high air pressure.

3.2.2. Air Tightness Time of Filter Cake

The air tightness time denotes the longest duration that the filter cake can maintain its air tightness under constant air pressure. The planned maximum operating time for pressurized chamber operations must not exceed the theoretical air tightness time of the filter cake. In this experiment, a constant air pressure value of 30 kPa was applied. The failure state of the filter cake after losing air tightness under constant air pressure is illustrated in Figure 9, primarily showing dry crack failure, occasionally accompanied by multiple circular puncture holes.

Figure 9. The form of failure of filter cake air tightness in constant air pressure.

The permeation flow rate curve of the filter cake under long-term air tightness until failure under constant air pressure is shown in Figure 10 (using the S1 strata as an example). Initially, the permeation flow rate remains close to zero and gradually increases over time. At a certain moment, there is a sudden increase in permeation flow rate, indicating the air tightness failure of the filter cake. Based on the permeation flow curves monitored during the air tightness process, the process can be divided into three stages, as shown in Figure 11. The first stage is the complete air tightness stage (AB). During a short period after pressurization, the filter cake hardly generates any permeation flow rate. The second stage is the filter cake compression stage (BC). It is in this state during most of the time, and the permeation flow curve approximates a straight line. The slope of the straight line is related to the pore structure of the filter cake, the filter cake material, and its water storage capacity. The third stage is the failure stage of air tightness. When reaching point C, water in a connected pore of the filter cake is completely displaced by air, and the air tightness of the filter cake fails, resulting in a sudden increase in permeation flow rate. During the process of air displacing pore water, cracks are generated on the air pressure side of the filter cake due to dehydration. As the displacement process progresses, the cracks continuously propagate inward, resulting in dry crack failure on the upper surface of the filter cake when the air tightness of the filter cake fails.

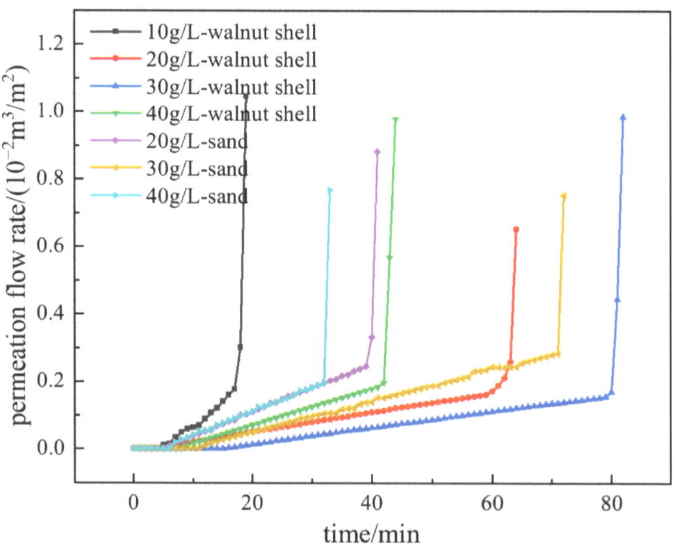

Figure 10. Flow volume through the filter cake under constant air pressure with varied coarse particle content.

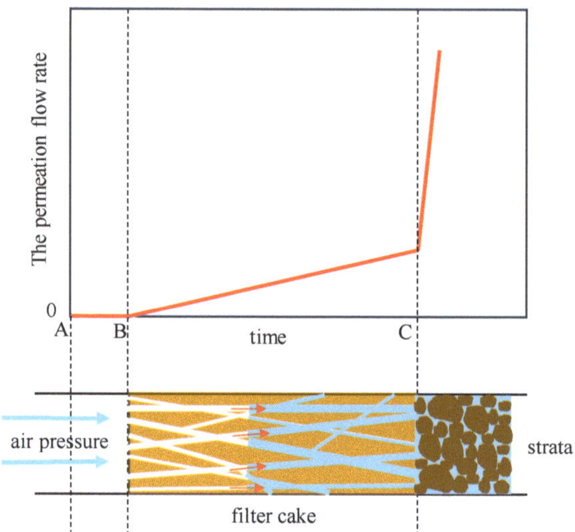

Figure 11. Three stages of air tightness of filter cake.

In the experiment, the time corresponding to point C, which generates a sudden increase in flow rate, is taken as the air tightness time. The air tightness time of each group of filter cakes is statistically analyzed, as shown in Figure 12, and it follows a similar trend to the air tightness pressure values. As the permeability coefficient of the strata increases, fewer filter cake test groups reach the air tightness pressure of 30 kPa, resulting in shorter air tightness times for the slurry. In the S1 and S2 strata, filter cakes with both types of coarse particles reach the longest air tightness time at a content of 30 g/L. Moreover, filter cakes with the same content of walnut shell particles have longer air tightness times than those with sand particles. In the S3 stratum, which has the highest permeability coefficient, filter cakes with both types of particles reach the longest air tightness time when the content is 40 g/L, and their maximum air tightness time is significantly shorter compared to the previous two strata.

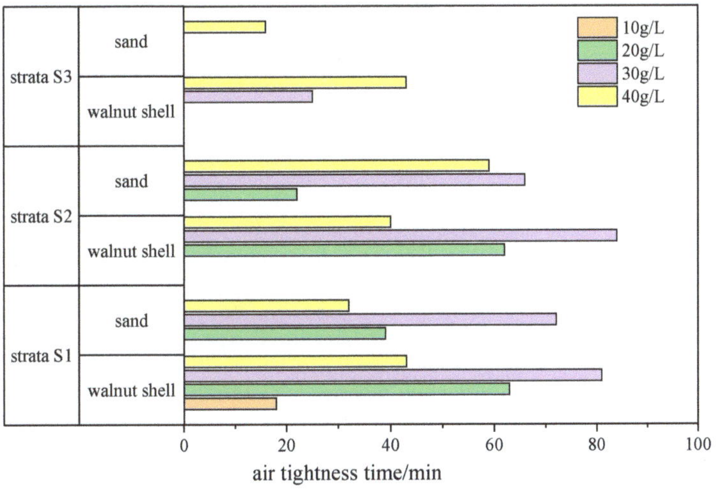

Figure 12. The air tightness time of the filter cake under varied permeability strata and coarse particle content.

4. Discussion

Experiment results indicate that both coarse particles play a significant role in the filter cake formation of drilling slurries in highly permeable strata. Currently, the coarse particles added to the slurry are all of a single particle size, with the selection of particle size focusing more on the pore characteristics of the strata while neglecting its compatibility with the base slurry. In reality, the formation of filter cakes in highly permeable strata requires the continuous blocking formed by the base slurry, coarse particles, and the strata. The coarse particles act as a "bridge" between the strata and the fine particles such as bentonite in the slurry. If the coarse particle size is too small, they will pass through the strata pores together with the slurry, failing to serve as a skeleton for plugging and bridging. Conversely, if the coarse particle size is too large, although they can form a bridge on the surface of the strata, the skeleton of the filter cake becomes too large, still allowing fine particles in the slurry to penetrate the strata through the pores of the coarse particle skeleton, resulting in the inability to form a dense filter cake. Therefore, it is necessary to simultaneously control the maximum and minimum particle sizes of the coarse particles to achieve effective blocking.

To achieve effective plugging, the particle size of the plugging material should be larger than the pore size of the medium. Thus, the interparticle pore size can be used to determine the particle size of the plugging material. The interparticle pore range for each particle group can be obtained using the triaxial pore calculation method proposed by Herzig et al. [16]. Taking the permeation test in the S3 strata as an example, the calculated interparticle pore range for the strata particles and different sizes of coarse particles are shown in Table 5. Since the maximum particle size of the base slurry is 0.075 mm, the minimum particle group of the coarse particles should be 0.25–0.5 mm. The interparticle pore range of the strata is 0.462–0.770 mm, and the maximum particle group of the coarse particles should be 0.5–1.0 mm (standard sieve pore size). Therefore, the coarse particle range for the S3 strata to meet the filter cake forming requirements should be 0.25–1.0 mm, which is consistent with the experimental results.

Table 5. Variation in pore size between particles.

Particle Size Range/mm	Interparticle Pore Range/mm
3–5	0.462–0.770
0.25–0.5	0.0385–0.0774
0.5–1	0.0774–0.154

Although the addition of both types of coarse particles effectively forms a filter cake, the filter cake containing walnut shell exhibited superior air tightness performance compared to the sand-containing filter cake under identical conditions. The reasons can be explained as follows. While sand-containing slurry often suffers from uneven filter cake formation due to the sedimentation effects of sand particles, walnut shell-containing slurry exhibits superior dispersion stability, facilitating the formation of a uniform and stable filter cake. As indicated by the basic properties of the slurry in Table 2, under identical conditions, the funnel viscosity of the walnut shell-containing fluid is higher than that of the sand-containing fluid, suggesting a more stable three-dimensional network structure in the former. Additionally, the ability of walnut shells to absorb water and undergo some degree of flexible deformation enhances the plugging of pore spaces in the formation, thereby promoting the formation of filter cake.

As the permeability coefficient of the strata increased, the optimal coarse particle content corresponding to the best air tightness performance of the filter cake gradually increased. The influence pattern of the content of both types of particles on air tightness performance was fundamentally consistent. Considering both the permeation flow rate of the filter cake and its air tightness performance, for high-permeability sand strata with permeability coefficients of 5.04×10^{-3} m/s and 5.07×10^{-2} m/s, it is recommended to use a coarse particle content of 30 g/L. For sand strata with a permeability coefficient of 2.30×10^{-1} m/s, a coarse particle content of 40 g/L is recommended.

5. Conclusions

To address the difficulty of filter cake formation and air tightness in high-permeability strata during the construction of slurry-water shield tunneling, comparative experiments were conducted using slurries containing two different coarse particles, walnut shells and sand, in three different permeability sand layers. The influence of coarse particles on filter cake formation and air tightness was analyzed, yielding the following main conclusions:

(1) The addition of coarse particles is essential for filter cake formation in high-permeability strata. The selection of coarse particle size should facilitate bridging effects. Controlling the maximum and minimum particle sizes based on the range of strata and coarse particle pore accumulation ensures that coarse particles can fill the strata pores while fine particles can fill the coarse particle pores.

(2) The permeation flow of the filter cake increases with increasing strata permeability, and firstly decreases and then increases with increasing particle addition. In vertical permeation, the permeation flow of walnut shell filter cake is greater than that of sand filter cake. The permeability coefficient of filter cakes with both coarse particles reaches less than 1.0×10^{-7} m/s, meeting the requirements of shield tunneling construction. The permeability coefficient of filter cake increases with increasing particle addition.

(3) The higher the strata permeability, the more difficult it is for the filter cake to achieve air tightness. The filter cake fails under high air pressure in the form of circular hole breakthrough damage and in the form of dry cracking damage under constant pressure. With increasing coarse particle addition, both the air tightness pressure value and air tightness time of the filter cake first increase and then decrease. Under the same conditions, filter cakes containing walnut shells exhibit a better air tightness performance than those containing sands.

(4) When performing pressurized maintenance operations with shield machines in high-permeability sand layers, using walnut shells as coarse particles instead of traditional

sand can improve the air tightness performance of filter cakes. In strata with a permeability coefficient less than 5.07×10^{-2} m/s, a recommended walnut shell content of 30 g/L is suggested. When the strata permeability coefficient is greater than 5.07×10^{-2} m/s, a recommended walnut shell content of 40 g/L is suggested.

Author Contributions: Conceptualization, Q.D. and T.L.; methodology, Y.W.; software, Q.D. and T.L.; validation, T.L. and S.L.; formal analysis, Y.W. and T.L.; investigation, T.L.; resources, T.L.; writing—original draft preparation, Q.D. and T.L.; writing—review and editing, Y.W. and L.W.; visualization, S.L.; supervision, Y.W., S.L. and L.W.; funding acquisition, Y.W., S.L. and L.W. All authors have read and agreed to the published version of the manuscript.

Funding: This research was funded by the National Natural Science Foundation of China, grant number 52309128; the National Natural Science Foundation of China, grant number U23B20144; the Research Project of China Railway 14th Bureau Group Co., Ltd., grant number 91370000163055989120106; and the Project funded by China South-to-North Water Diversion Middle Route Co., Ltd., grant number ZXJ/KJ/YW-025.

Institutional Review Board Statement: Not applicable.

Informed Consent Statement: Not applicable.

Data Availability Statement: The original contributions presented in the study are included in the article, further inquiries can be directed to the corresponding author.

Conflicts of Interest: Author Liu Sijin was employed by the China Railway 14th Bureau Group Co., Ltd. and author Wen Letian was employed by the China South-to-North Water Diversion Middle Route Co., Ltd. The remaining authors declare that the research was conducted in the absence of any commercial or financial relationships that could be construed as a potential conflict of interest.

References

1. Morris, M.; Yang, M.W.; Tsang, C.K.; Hu, A.Y.; Shut, D.S. *An Overview of Subsea Tunnel Engineering in Hong Kong*; Thomas Telford Ltd.: London, UK, 2016.
2. Lin, C.; Zhang, Z.; Wu, S.; Yu, F. Key techniques and important issues for slurry shield under-passing embankments: A case study of Hangzhou Qiantang River Tunnel. *Tunn. Undergr. Space Technol.* **2013**, *38*, 306–325. [CrossRef]
3. Min, F.; Song, H.; Zhang, N. Experimental study on fluid properties of slurry and its influence on slurry infiltration in sand stratum. *Appl. Clay Sci.* **2018**, *161*, 64–69. [CrossRef]
4. Naghadehi, M.Z.; Thewes, M.; Lavasan, A.A. Face stability analysis of mechanized shield tunneling: An objective systems approach to the problem. *Eng. Geol.* **2019**, *262*, 105307. [CrossRef]
5. Liu, K.; Ding, W.; Qu, C. A Mesoscopic Viewpoint on Slurry Penetration and Pressure Transfer Mechanisms for Slurry Shield Tunneling. *Buildings* **2022**, *12*, 1744. [CrossRef]
6. Talmon, A.M.; Mastbergen, D.R.; Huisman, M. Invasion of Pressurized Clay Suspensions into Granular Soil. *J. Porous Media* **2013**, *16*, 351–365. [CrossRef]
7. Kou, L.; Zhao, J.; Miao, R.; Lian, F. Experimental Study on the Formation and Characteristics of Mud Filtration Cake in Large-Diameter Slurry Shield Tunneling. *Adv. Civ. Eng.* **2021**, *2021*, 1–10. [CrossRef]
8. Liu, D.; Liu, X.; Lin, C.; Xiong, F.; Han, Y.; Meng, Q.; Zhong, Z.; Chen, Q.; Weng, C. Experimental study and engineering application of slurry permeability mechanism of slurry shield in circular-gravel stratum. *Arab. J. Geosci.* **2020**, *13*, 1000. [CrossRef]
9. Jin, D.; Shen, Z.; Song, X.; Yuan, D.; Mao, J. Numerical Analysis of Slurry Penetration and Filter Cake Formation in Front of Tunnel Face. *Tunneling Undergr. Space Technol.* **2023**, *140*, 105303. [CrossRef]
10. Fritz, P. Additives for Slurry Shields in Highly Permeable Ground. *Rock Mech. Rock Eng.* **2007**, *40*, 81–95. [CrossRef]
11. Wang, Z.; Wang, Y.; Feng, D.; Zhang, J.; Liu, S. Effects of slurry viscosity and particle additive size on filter cake formation in highly permeable sand. *Undergr. Space* **2022**, *7*, 151–161. [CrossRef]
12. Lei, H.; Liu, X.; Shi, F.; Ma, C. Infiltration behaviour into sand of bentonite slurry with guar gum. *Géotechnique Lett.* **2022**, *12*, 161–166. [CrossRef]
13. Wang, Z.; Guo, W.; Qin, W.; Wang, C.; Ding, W. Influences of Polyanionic Celluloses and Temperature on the Rheological Property of Seawater Slurries. *Constr. Build. Mater.* **2022**, *351*, 128964. [CrossRef]
14. Ruth, B.F. Studies in filtration III. Derivation of general filtration equations. *Ind. Eng. Chem.* **1935**, *27*, 708–723. [CrossRef]
15. Tien, C.; Bai, R. An assessment of the conventional cake filtration theory. *Chem. Eng. Sci.* **2003**, *58*, 1323–1336. [CrossRef]
16. Herzig, J.P.; Leclerc, D.M.; Le Goff, P. *Flow of Suspensions Through Porous Media: Application to Deep Filtration*; American Chemical Society: Washington, DC, USA, 1970.
17. Van Oort, E.; van Velzen, J.F.G.; Leerlooljer, K. Impairment by Suspended Solids Invasion Testing and Prediction. *Spe Prod. Facil.* **1993**, *8*, 178–184. [CrossRef]

18. Min, F.; Zhu, W.; Han, X. Filter Cake Formation for Slurry Shield Tunneling in Highly Permeable. *Tunn. Undergr. Space Technol.* **2013**, *38*, 423–430. [CrossRef]

19. Xu, T.; Bezuijen, A. Bentonite slurry infiltration into sand: Filter cake formation under various conditions. *Géotechnique* **2019**, *69*, 1095–1106. [CrossRef]

20. Qin, S.; Xu, T.; Zhou, W.H.; Bezuijen, A. Infiltration Behaviour and Microstructure of Filter Cake from Sand-modified Bentonite Slurry. *Transp. Geotech.* **2023**, *40*, 100963. [CrossRef]

21. Wang, Y.; Wang, Z.; Feng, D.; Liu, S.; Xu, S.; Zhou, K. Seepage Characteristics of Slurry with Particle Additives in the High-Permeability Sand Layer. *Geofluids* **2021**, *2021*, 5546932. [CrossRef]

22. Lin, Y.; Fang, Y.; He, C. Investigations of filter-clog mechanism and prediction model of slurry penetration during slurry pressure transfer. *Acta Geotech.* **2023**, *18*, 5251–5272. [CrossRef]

23. Ma, T.; Peng, N.; Chen, P. Filter cake formation process by involving the influence of solid particle size distribution in drilling fluids. *J. Nat. Gas Sci. Eng.* **2020**, *79*, 103350. [CrossRef]

24. Ike, D.C.; Ibezim-Ezeani, M.U.; Akaranta, O. Cashew nutshell liquid and its derivatives in oil field applications: An update. *Green Chem. Lett. Rev.* **2021**, *14*, 620–633. [CrossRef]

25. Agi, A.; Oseh, J.O.; Gbadamosi, A.; Fung, C.K.; Junin, R.; Jaafar, M.Z. Performance evaluation of nanosilica derived from agro-waste as lost circulation agent in water-based mud. *Pet. Res.* **2023**, *8*, 256–269. [CrossRef]

26. Kai, C.; Zhang, F.J.; Cheng, C.L.; Chen, Q.B. Design synthesis and performance of anti-collapse drilling polymer mud with higher stability. *Pigment. Resin Technol.* **2022**, *51*, 101–109. [CrossRef]

27. Ebrahimi, M.A.; Sanati, A. On the potential of alyssum as an herbal fiber to improve the filtration and rheological characteristics of water-based drilling muds. *Petroleum* **2022**, *8*, 509–515. [CrossRef]

28. Elahifar, B.; Hosseini, E. Laboratory study of plugging mechanism and seal integrity in fractured formations using a new blend of lost circulation materials. *J. Pet. Explor. Prod. Technol.* **2023**, *13*, 1197–1234. [CrossRef]

29. Ab Lah, N.N.; Ngah, K.; Sauki, A. Study on the viability of egg shell as a lost circulation material in synthetic based drilling fluid. *J. Phys. Conf. Ser.* **2019**, *1349*, 12135.

30. Ali, J.A.; Abdalqadir, M.; Najat, D.; Hussein, R.; Jaf, P.T. Application of ultra-fine particles of potato as eco-friendly green additives for drilling a borehole: A filtration, rheological and morphological evaluation. *Chem. Eng. Res. Des.* **2024**, *206*, 89–107. [CrossRef]

31. Li, X.; Yuan, D. Creating a working space for modifying and maintaining the cutterhead of a large-diameter slurry shield: A case study of Beijing railway tunnel construction. *Tunn. Undergr. Space Technol.* **2018**, *72*, 73–83. [CrossRef]

32. Herrenknecht, M.; Bäppler, K. Compressed air work with tunnel boring machines. In *Underground Space–The 4th Dimension of Metropolises*; Routledge: Oxfordshire, UK, 2007.

33. Kim, S.H.; Tonon, F. Face stability and required support pressure for TBM driven tunnels with ideal face membrane–Drained case. *Tunn. Undergr. Space Technol.* **2010**, *25*, 526–542. [CrossRef]

34. Min, F.; Liu, J.; Chen, J.; Liu, T.; Yu, C.; Ji, J.; Liu, J. A study on the excavation face failure of pressurized slurry shield. *Tunn. Undergr. Space Technol.* **2023**, *132*, 104900. [CrossRef]

35. Zhang, S.; Ding, J.; Jiao, N.; Liu, J.; Sun, S. Destruction Mechanism of Filter Cake under Air Pressure during Chamber Opening. *J. Test. Eval.* **2024**, *52*, 897–914. [CrossRef]

36. Fredlund, D.G.; Rahardjo, H. *Soil Mechanics for Unsaturated Soils*; Wiley: Hoboken, NJ, USA, 1993.

37. Lee, C.; Wu, B.R.; Chen, H.T.; Chiang, K.H. Tunnel stability and arching effects during tunneling in soft clayey soil. *Tunn. Undergr. Space Technol.* **2006**, *21*, 119–132. [CrossRef]

38. Min, F.; Zhu, W.; Xia, S.; Wang, R.; Wei, D.; Jiang, T. Test Study on Airtight Capability of Filter Cakes for Slurry Shield and Its Application in a Case. *Adv. Mater. Sci. Eng.* **2014**, *2014*, 696801. [CrossRef]

39. Xia, S. *Study on Airtightness Mechanism of Filter Cake under Compressed Air in Slurry Shield*; Hohai University: Nanjing, China, 2012.

40. Cai, B.; Jin, D.; Li, X. Theory of Failure of Filter Cake Airtightness During Chamber Opening in a Pressurized Shield. *Rock Soil Mech.* **2023**, *44*, 1395–1415.

41. Liu, C.; Guo, J.; Yang, L.; Gao, Y. Effects of Coarse Material in Slurry on the Airtightness Time for the Filter Cake in Slurry Shield Tunneling. *China Sci.* **2017**, *12*, 1508–1513.

42. Hilal, N.; Ali, T.K.M.; Tayeh, B.A. Properties of environmental concrete that contains crushed walnut shell as partial replacement for aggregates. *Arab. J. Geosci.* **2020**, *13*, 812. [CrossRef]

materials

Article

Bond Strength Assessment of Normal Strength Concrete–Ultra-High-Performance Fiber Reinforced Concrete Using Repeated Drop-Weight Impact Test: Experimental and Machine Learning Technique

Sadi I. Haruna [1,*], **Yasser E. Ibrahim** [1], **Ibrahim Hayatu Hassan** [2], **Ali Al-shawafi** [3,*] **and Han Zhu** [1]

[1] Engineering Management Department, College of Engineering, Prince Sultan University, Riyadh 11586, Saudi Arabia; ymansour@psu.edu.sa (Y.E.I.); hanzhu2000@tju.edu.cn (H.Z.)
[2] Institute of Agricultural Research, Ahmadu Bello University, Zaria 810006, Nigeria; ihhassan@abu.edu.ng
[3] School of Civil Engineering, Tianjin University, Tianjin 300350, China
* Correspondence: sharuna@psu.edu.sa (S.I.H.); ali91@tju.edu.cn (A.A.-s.)

Abstract: Ultra-high-performance concrete (UHPC) has been used in building joints due to its increased strength, crack resistance, and durability, serving as a repair material. However, efficient repair depends on whether the interfacial substrate can provide adequate bond strength under various loading scenarios. The objective of this study is to investigate the bonding behavior of composite U-shaped normal strength concrete–ultra-high-performance fiber reinforced concrete (NSC-UHPFRC) specimens using multiple drop-weight impact testing techniques. The composite interface was treated using grooving (Gst), natural fracture (Nst), and smoothing (Sst) techniques. Ensemble machine learning (ML) algorithms comprising XGBoost and CatBoost, support vector machine (SVM), and generalized linear machine (GLM) were employed to train and test the simulation dataset to forecast the impact failure strength ($N2$) composite U-shaped NSC-UHPFRC specimen. The results indicate that the reference NSC samples had the highest impact strength and surface treatment played a substantial role in ensuring the adequate bond strength of NSC-UHPFRC. NSC-UHPFRC-Nst can provide sufficient bond strength at the interface, resulting in a monolithic structure that can resist repeated drop-weight impact loads. NSC-UHPFRC-Sst and NSC-UHPFRC-Gst exhibit significant reductions in impact strength properties. The ensemble ML correctly predicts the failure strength of the NSC-UHPFRC composite. The XGBoost ensemble model gave coefficient of determination (R^2) values of approximately 0.99 and 0.9643 at the training and testing stages. The highest predictions were obtained using the GLM model, with an R^2 value of 0.9805 at the testing stage.

Keywords: normal strength concrete; UHPFRC; impact strength; surface treatment; machine learning algorithms

Citation: Haruna, S.I.; Ibrahim, Y.E.; Hassan, I.H.; Al-shawafi, A.; Zhu, H. Bond Strength Assessment of Normal Strength Concrete–Ultra-High-Performance Fiber Reinforced Concrete Using Repeated Drop-Weight Impact Test: Experimental and Machine Learning Technique. *Materials* **2024**, *17*, 3032. https://doi.org/10.3390/ma17123032

Academic Editors: Oldrich Sucharda and Francesco Fabbrocino

Received: 26 April 2024
Revised: 29 May 2024
Accepted: 17 June 2024
Published: 20 June 2024

1. Introduction

Because of its superior compressive strength and durability, UHPC is becoming increasingly popular in many building construction industries. These ideal UHPC characteristics can decrease the size and self-weight of a structure remarkably. UHPC is frequently used as a joint to combine concrete members [1,2] and a repair and retrofit material for damaged concrete structures [1]. UHPC has been used to enhance and strengthen concrete elements, including bridges and decks; this increases the behavior of the existing reinforced concrete structures, including hardened and durability properties. The NSC-UHPC composites involve the interface between the substrate and new cementitious materials with weak areas [1,2]. The bonding behavior at the interface considerably affects the overall performance of the concrete structure [3]. The interface of NSC-UHPC composites is under a self-equilibrated state of tension collective with shear [4].

The normal stress states at the interface, such as tension, shear, and shear–compression combinations, are essential. Therefore, many studies have investigated and modeled the bond behavior between old and new cementitious materials using different laboratory methods, including (i) tension tests such as pull-off, direct tension, and splitting, and (ii) shear tests that include direct shear and slant shear. The experiments described above were performed under static loading conditions. The interfacial bond of composite NSC-UHPC is primarily examined under static load application. However, UHPC material is widely used to repair concrete structures such as military structures, bridge columns, and road and runway facilities that are usually exposed to impact loads. Studying the bond performance of NSC-UHPC composite under impact loads can aid in selecting the best material for NC structural repair and retrofitting application.

Yu et al. [5] investigated the debonding failure pattern of the NSC-UHPC composite using a multi-scale technique. The results revealed that the NC substrate's interface roughness and strength increased mechanical performance. Yang et al. [6] examined the bending behavior of a precast NSC-UHPC specimen. The results showed that the hybrid NSC-UHPC exhibits sufficient capacity, stiffness, etc., and the composite section displayed large fracture widths. Jongvivatsakul et al. [7] reported improved bond strength between carbon-fiber-reinforced polymer plates and concrete using carbon nanotube-reinforced epoxy composites. The U-shaped drop-weight impact testing approach is a newly developed impact testing technique used to evaluate the impact properties of cementitious materials [8–10]. The approach evaluated impact strength using a unique U-shaped sample and a repeated drop weight. In addition, the types and impact properties of UHPC under diverse loading circumstances were examined in past studies, including low-velocity impact [11], projectile loads [12,13], dynamic splitting [14], dynamic mechanical method [15], and multiple impact loads [16]. Yu et al. [17] used pendulum loads to determine the impact strength of UHPFRC samples. They discovered that the fiber length is the primary cause of the increased energy dissipation capacity in the UHPC.

The traditional approaches for forecasting these features are based on empirical interactions established from broad lab research. Conversely, the complication and non-linear parameters determining the impact resistance of concrete require advanced prediction techniques. The capability of ML revealed the remarkable capability to handle complicated relationships in a broad database, which demonstrates a pronounced perspective in several engineering fields. Almustafa and Nehdi [18] developed an ensemble tree-based model for estimating the structural behavior of reinforced concrete (RC) columns under blast loads. The authors predicted the blast characteristics of individual elements of RC structures to eventually construct an adaptable model for the global performance of the complete RC structures. Zhang et al. [13] predicted the impact and blast strength of UHPC using the Karagozian and Case (K&C) material model. The technique's prediction ability was demonstrated by apprehending numerous experimental datasets. Cao et al. [19] forecasted the impact strength of concrete containing fibers using the adaptive neuro-fuzzy inference system (ANFIS) model. The UHPC's dynamic mechanical characteristics were assessed using the ML model [20]. Shao et al. [21] evaluated the UHPC's penetrating resistance encapsulated in a ceramic ball exposed to projectile stress. However, little research has investigated the impact behavior of UHPFRC using repeated drop-weight impact tests and an artificial intelligence-based algorithm to analyze impact resistance. Therefore, the objective of this study is to assess the bonding strength of the NSC-UHPFRC composite under multiple drop-weight impact tests. The composite NSC-UHPFRC's interface was treated using three surface treatment methods, which include normal fracture (Nst), smooth (Sst), and grooves surface (Gst). Moreover, two ensemble models, Catboost and XGboost, and classical models (SVR and GLM models) were used to predict the impact strength ($N2$) of the U-shaped NSC-UHPFRC composite. The study explored the suitability of studying the bond strength properties of composite materials under impact stress.

2. Materials and Methods

2.1. Materials

2.1.1. NSC and UHPFRC

Table 1 presents the mixed proportion of the NSC and UHPFRC materials used in this study. Ordinary Portland cement (OPC) grade 52.5, confirmed with cement composition requirements specified in GB175-2007 [22]. The NSC mixture was prepared following Chinese standard JTG55-2011 [23]; the NSC mixture was designed to achieve a compressive strength of 40 MPa. The NSC constituent materials include natural river sand as fine aggregate and crushed stone as medium aggregate (10 mm particle size). At the same time, the UHPFRC mixture was prepared according to GB175-2007 [22], designed to achieve a compressive strength of 125 MPa. The constituent material of UHPC includes OPC, silica fumes (SF), slag, quartz powder was bought from Zhixiang Industrial Trading Co., Ltd. (Yongdeng, China), and a water-reducing agent was sourced from Jiangsu Subot New Materials Co., Ltd. (Nanjing, China),. The performance parameters of the steel fiber manufactured by Ganzhou Daye Metal Fiber Co., Ltd. (Ganzhou, China) was added to the UHPC mixture by volume fraction as summarized in Table 2.

Table 1. Mixture of the proportion of NSC and UHPFRC (kg/m^3).

Materials	NSC	UHPFRC
OPC	420	1000
Fine aggregate	573	1200
Medium aggregate	1273	0.00
Water	185	232
Quartz powder	0.00	50.0
Water-reducing agent	0.63	200
Slag	0.00	200
Silica fumes	0.00	250
Steel Fiber (V_f%)	0.00	1.0

Table 2. Technical indexes of micro steel fiber.

Properties	Length/mm	Diameter/mm	Aspect Ratio	Density/kg/m^3	Tensile Strength/MPa
	13.0	0.2	65.0	7800	2850

2.1.2. Sample Fabrications

This study fabricated U-shape cubes (100 × 100 × 100 mm^3) and beam composite specimens (100 × 100 × 400 mm^3) by bonding half the NSC substrate and half the UHPFRC layer together to form composite NSC-UHPFRC specimens. The preparation procedures for the production of the NSC-UHPFRC composite are illustrated in Figure 1. Three surface treatment methods were adopted at the interface based on the natural fracture, smooth, and grooves surfaces (see Figure 2). For each specimen condition, twenty (20) control and composite NSC-UHPFRC specimens were fabricated and tested for multiple drop-weight impacts to assess the bonding behavior at the interface. The specimens treated with different surface interfaces were produced after casting the NSC U-shaped specimen in the mold and cured for 28 d. The cured samples, including the cube and beams, were then cut into two pieces considering smoothed, natural fracture, and grooved surfaces. The NSC substrate surface was treated using longitudinal grooving and smoothing techniques. The last group was made by just breaking the NSC samples naturally. The half pieces for each surface treatment were then put back into the mold to cast the remaining half of the UHPFRC. After casting the composite NSC-UHPFRC, the samples were then kept at a standard curing room temperature for another 28 d for the UPHFRC part to attain its maximum before testing.

Figure 1. Schematic procedure for the production of NSC-UHPFRC specimens.

Figure 2. Surface treatment techniques adopted in this study.

2.2. Testing Methods

Previous studies [24] have indicated many testing methods for investigating the bond behavior at the interface of composite materials. These are classified into two: (I) testing under the synergy effect of shear and compression stress, for instance, the most reliable bond strength result may not be determined using the slant shear test due to the existence of compression force. (II) Testing under pure sheer stress includes splitting tensile, 3-point flexural test, direct shear, etc. This research attempted to evaluate the bond behavior of the NSC-UHPFRC composite at the interface under multiple drop-weight impact testing techniques.

Multiple Drop-Weight Impact Tests

Figure 3 presents the experimental process for impact tests. The NSC-UHPFRC specimens were exposed to repeated impact loads using a hammer weighing 2.1 kg, which was released from a 457 mm distance. The impact test was carried out following a modified version of the ACI 544- test approach [25]. The impact loads that induced the first crack and failure strength were recorded as $N1$ and $N2$, respectively, reflecting the bond resistance subjected to dynamic load. The absorbed impact energy is determined using Equations (1) and (2) for impact mass (w), acceleration due to gravity (g), falling mass velocity (v), from height (h) at the first and complete failure stages, respectively. The test specimens are instrumented with strain gauges to enable the detection of the occurrence of the first crack, which is transmitted to the data acquisition system.

$$IE_1 = N_1.wgh = N_1 \frac{wv^2}{2} \tag{1}$$

$$IE_2 = N_2.wgh = N_2 \frac{wv^2}{2} \tag{2}$$

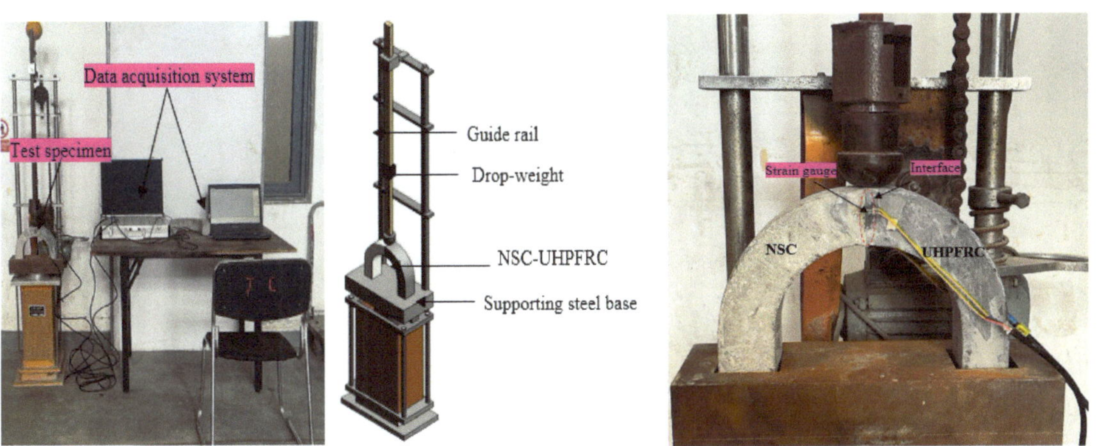

Figure 3. Multiple drop-weight Impact tests.

2.3. Machine Learning Algorithms

In this paper, two ensemble machine learning models (XGboost and CatBoost), SVM and GLM, were used to forecast $N2$ at the fracture stage of the test specimen under multiple drop-weight impact tests python version 3.10 The developed models were used to train the datasets acquired from this study and previous literature [26]. To model $N2$, sensitivity analysis was used to select the input variables in the XGBoost, CatBoost, SVR, and GLM models, as depicted in Equation (3)

$$failure\ strength\ (N_2\ (blows)) = \begin{cases} XGBoost = f(Cfc + P + \rho + N_1) \\ CatBoost = f(Cfc + P + \rho + N_1) \\ SVR = f(Cfc + P + \rho + N_1) \\ GLM = f(Cfc + P + \rho + N_1) \end{cases} \tag{3}$$

where f defines the function of the input variable, Cfc is the compressive strength of the composite, P = flexural load, ρ is the density, and $N1$ is the first crack strength.

The modeling dataset was divided into a ratio of 70:30 for the training and testing stages. The efficiency of the two models was validated using a 10-fold cross-validation approach. This approach is reported in the literature [27] as the best method for providing unbiased model evaluation for a small dataset. The dataset was also normalized using Equation (4) to enhance its integrity and decrease redundancy.

$$x_n = \frac{x - x_{min}}{x_{max} - x_{min}} \tag{4}$$

where x_n is the scaled value, x is the un-normalized value, x_{min} is the smallest value, and x_{max} defines the largest value in the dataset.

2.3.1. CatBoost

CatBoost, referred to as categorical boosting, is an enhanced ML technique based on the gradient-boosting tree structure. It is built on the symmetric decision tree technique and largely handles the issues of efficiently managing categorical information, gradient bias, and estimate offsets. CatBoost tries to improve the algorithm's skills and general capabilities by combining the results of all trees to get the final solution [28]. CatBoost generates plus one independent random sample permutations from a given training set. The main objective of the CatBoost implementation was to tackle prediction shift issues. Manual settings were considered when choosing the hyperparameters for the experiment. The iterations were set to 50,000, the maximum depth to 3, the learning rate to 0.1, the loss function to "RMSE, MSE, MAE, and R^2", and other default settings were maintained at default values.

2.3.2. XGBoost

XGBoost is a robust supervised tree-based ML model widely employed in several disciplines because of its high accuracy and ability to overcome overfitting difficulties commonly encountered in other tree-based approaches [29,30]. XGBoost was built using the Gradient Boosting technique, which improves performance by mixing weak models. This technique processes the input parameters and then builds a model to implement the following tasks: ranking, regression, and classification. Here, we engaged the XGBoost to handle regressive prediction tasks. A compressed mathematical design of the loss function using the XGBoost hyperparameters is depicted in Equation (5).

$$L = \sum_{j=1}^{T_m} \left[G_{jm} W_{jm} + \frac{1}{2} (H_{jm} + \lambda_R) W^2_{jm} + \alpha_R |W_{jm}| \right] + \gamma T_m \tag{5}$$

where L is the loss function, α_R and λ_R indicate $L1$ and $L2$ regularization factors, respectively, and γ designates the penalization variable. H_{jm} and G_{jm} signify the sum of hessian and the sum of gradient, respectively, for the optimum weight for each region, j (H_{jm} was applied to obtain the smallest sum of child weights) and W_{jm} represents the optimum weights. T_m denotes leave tree over a maximum depth of tree D_{max}.

2.3.3. Support Vector Regression (SVR)

SVR is an adaptation of the SVM and has been successfully applied to regression evaluation in a number of scientific and technical applications. The primary goal of an SVR model is to accurately estimate the output parameter, $\{p_i\}$, in accordance with a set of input parameters, $\{y_i\}$, by fitting a regression function, $P = f(y)$, correctly. The training dataset is represented by the expression $M = (y_1, p_1), (y_2, p_2), \ldots, (y_k, p_k)$, where $p_i \in R_K$ denotes the desired quantity and $y_i \in R_K$ is a vector representing the input features. For non-linearly connected data to the preferred output variables, as is the case in nonlinear scenarios observed in the real world, a nonlinear mapping function $x_i(y)$ as given in Equation (6) [31] can be used to build the linear relation in the high-dimensional feature space. The structure of the SVR model is shown in Figure 4.

$$f(y) = \sum_{i=1}^{n} w_i x_i(y) + b \tag{6}$$

where b is the bias, $w_i x_i(y)$ is the function referred to as the feature, and the dot product in the feature space is F. It is based on the structural risk minimization theory [32].

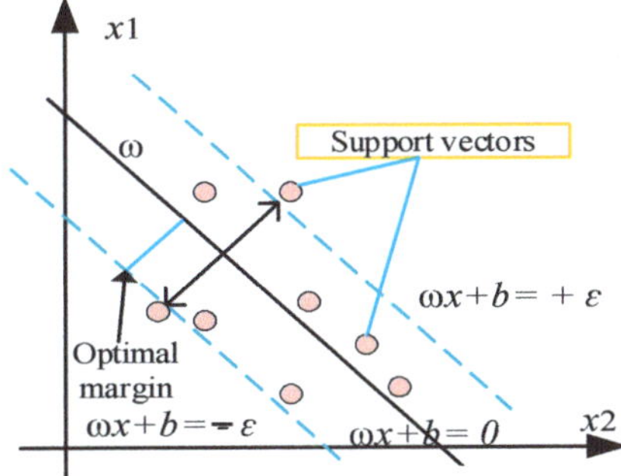

Figure 4. Structure of SVM.

2.3.4. Generalized Linear Model (GLM)

The GLM is frequently applied to binary or count data modeling [33]. GLM can be considered a nonlinear regression model. We built a predictive model using GLM and a Gamma (Γ) distribution. GLM can be characterized as nonlinear due to the characteristics of the CRI distribution in the lexicon of used data. An explanation of the GLM formulation is as follows:

$$y_i = \sum_{k=1}^{K} \beta_k x_{ki} + \beta_0 \tag{7}$$

where a linear function of the explanatory variable (input) x_{ki} with input coefficients β_k and bias value β_0 models the response y_i. The expectation of the dependent variable $\mu_i = E(y_i)$ is operated upon by an invertible link function $g(\mu_i.)$ to a linear predictor (y_i).

$$g(\mu_i) = y_i = \sum_{k=1}^{K} \beta_k x_{ki} + \beta_0 \tag{8}$$

2.3.5. Performance Evaluation Measures

To evaluate the efficiency of the ML models, four performance measures were used, including coefficient of determination (R^2), mean absolute error (MAE), mean square error (MSE), and root mean square error (RMSE). The expression of the matrix is described by Equations (9) to (11). Therefore, the accuracy of the model noted in the indicators was used to determine which machine learning models were the best. The modeling process is depicted in Figure 5.

$$R^2 = \left[\frac{\sum_{i=1}^{n} (q_i - \bar{q})(p_i - \bar{p})}{\sum_{i=1}^{n} (q_i - \bar{q})^2 \sum_{i=1}^{n} (p_i - \bar{p})^2} \right] \tag{9}$$

$$MAE = \frac{1}{2}\sum_{i=1}^{n}|q_i - p_i| \qquad (10)$$

$$MSE = \frac{1}{n}\sum_{i=1}^{n}(q_i - \overline{q_i})^2 \qquad (11)$$

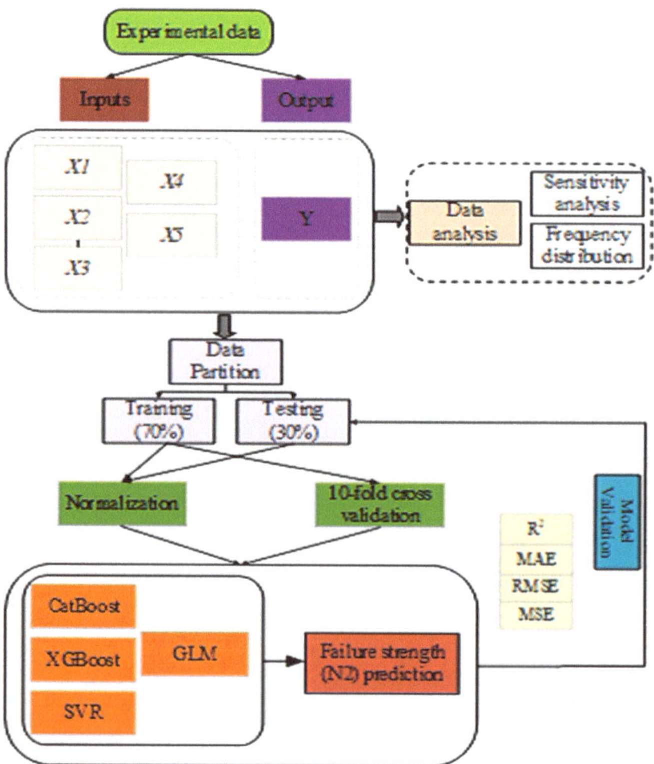

Figure 5. Flow chart of the model development.

3. Results and Discussions

3.1. Impact Resistance of NSC-UHPFRC

Figures 6 and 7 present the impact properties of NSC-UHPFRC composites under different testing conditions. The interface of composite NSC-UHPFRC specimens was treated with three different surface treatments. The effects of these surface treatments assess the composite bond's ability to resist repeated impact loads. The number of drops that caused the first crack strength ($N1$), failure strength ($N2$), and equivalent absorbed impact energy for the four testing conditions are shown in Figures 6 and 7, respectively. Equations (1) and (2) were applied to determine the absorbed impact energy as the two cracking stages were calculated based on drop weight and falling height. From Figure 7, it can be noted that the reference NSC specimen demonstrates a high drop number before the initial crack ($N1$) occurrence, with the average number of drops equal to 24 blows, followed by the specimen treated with natural fracture surface (NSC-UHPC-Nst). The U-shaped NSC-UHPFRC-Gst specimens reveal an average number of drops to initial the first crack strength of 11 blows, 118% lower than that of the control sample. The U-shaped composite formed with a smooth surface at the interface (NSC-UHPFRC-Sst) showed the lowest ability to resist the impact load at both the first and failure stages. The mean number of drops ($N1$) of the control NSC is 608.9% higher than that of the NSC-UHPFRC-Sst specimen. Similarly, the ability of the U-shaped composite to resist impact load caused complete failure of the specimens, which followed a similar pattern as the first crack strength. Reference NSC had the peak mean number of drops that caused complete failure, followed by NSC-UHPFRC-Nst, and then NSC-UHPFRC-Gst; the lowest average number of blows was recorded against NSC-UHPFRC-Sst specimens (see Figure 6). This finding shows that surface treatment plays a significant role in ensuring sufficient bond strength at the interface of the NSC-UHPFRC composite. The bond behavior between the NSC substrate under natural fracture and the UHPFRC layer can provide sufficient bond strength at the interface, resulting in a monolithic structure that can withstand dynamic loads under impact loads, as a slight decrease in the impact strength was obtained in the NSC-UHPFRC-Nst specimen compared to the reference NSC specimen.

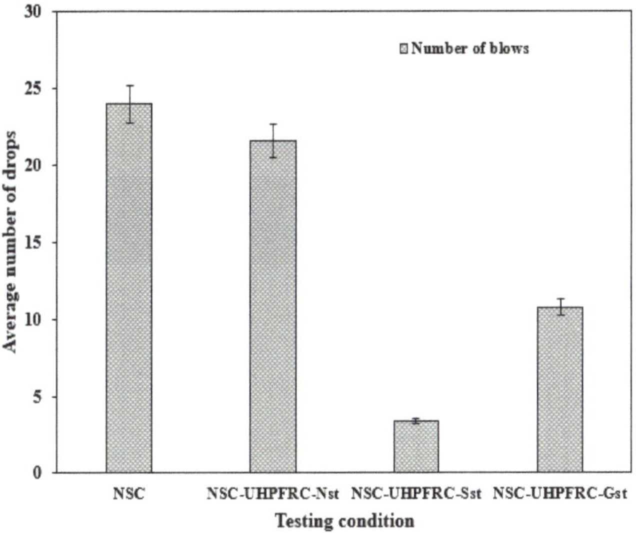

Figure 6. The impact strength ($N1$) of NSC-UHPFRC specimens.

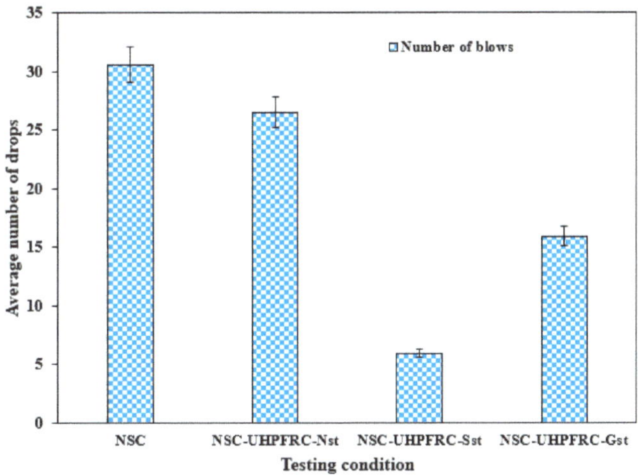

Figure 7. The impact strength ($N2$) of NSC-UHPFRC specimens.

The ductility index of the test specimen is defined as the ratio of the post-crack strength $(N2-N1)$ to the first-crack strength $(N1)$, which explains the U-shaped NSC-UHPFRC specimen's toughness after cracking [34,35]. The expression is given in Equation (9)

$$Ductility\ index\ (\eta) = \frac{(N2 - N1)}{N1} \tag{12}$$

Figure 8a presents the average ductility index (DI) value obtained under each testing condition, comprised of twenty (20) specimens demonstrating the ability of NSC-UHPFRC to absorb kinetic energy [34,36]. Because of the presence of SF in the UHPFRC, the ductility of the composite U-shaped was improved, which transformed the composite specimens to a more ductile state and enhanced the impact strength of the NSC-UHPFRC. The DI values of the NSC-UHPFRC for each testing condition are less than unity. The highest average DI value of 0.85 was obtained for NSC-UHPFRC-Sst specimens, attributed to the higher post-crack strength obtained in this group. The reference NSC, NSC-UHPFRC-Nst, and NSC-UHPFRC-Gst specimens had DI values of 0.29, 0.25, and 0.52, respectively. Figure 8b shows the COV of the impact strength data obtained at the two cracking phases ($N1$ and $N2$). As presented in Figure 8b, the NSC-UHPFRC-Sst specimen revealed the highest coefficient of variation of 35.29% at the initial crack point, with the lowest value of 19.51% at the failure stage ($N2$) among all testing groups. Interestingly, the maximum COV obtained in this work is lesser than the value achieved in many past studies [37,38] that utilized the drop-weight impact testing approach ACI 544-2R [25]. The reduction in COV is due to adopting a U-shape, which agrees with previous studies [8,39]. The overall impact test results are summarized in Table 3.

Table 3. Impact test results of composite NSC-UHPFRC.

S/N	NSC		NSC-UHPFRC-Nst		NSC-UHPFRC-Sst		NSC-UHPFRC-Gst	
	$N1$	$N2$	$N1$	$N2$	$N1$	$N2$	$N1$	$N2$
1	24	28	22	26	3	5	10	14
2	27	34	23	30	7	9	12	19
3	12	17	10	15	4	6	5	10
4	25	33	28	32	3	5	11	19
5	23	32	20	26	5	8	10	16
6	28	34	25	30	2	5	12	17
7	26	35	23	27	3	6	11	15
8	20	25	17	22	2	5	8	13
9	15	24	12	16	3	6	6	10
10	23	29	21	26	3	5	10	15
11	37	46	33	37	4	6	17	21
12	17	25	15	19	3	5	9	13
13	18	23	16	21	3	5	7	12
14	24	30	21	27	2	5	10	16
15	29	36	24	29	3	6	13	18
16	25	31	23	29	4	6	11	17
17	34	39	28	33	3	6	15	20
18	32	36	27	31	2	5	14	18
19	29	34	25	31	4	7	15	22
20	14	21	19	23	5	8	9	14
Mean	24.10	30.60	21.60	26.50	3.40	5.95	10.75	15.95
SD	6.49	6.63	5.52	5.58	1.20	1.16	3.00	3.32
COV.	26.92	21.66	25.54	21.06	35.29	19.51	27.89	20.84

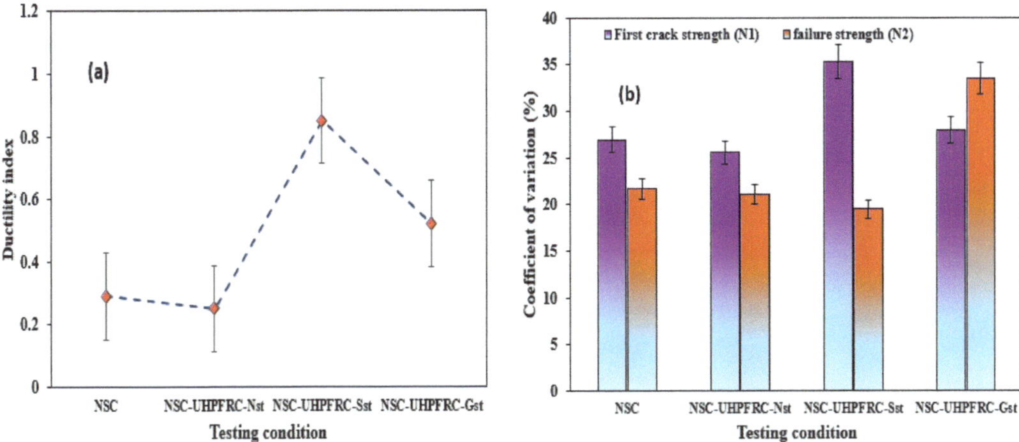

Figure 8. (**a**) Ductility index and (**b**) COV of the impact strength data.

3.2. NSC

The control U-shaped NSC specimens showed maximum impact strength compared to the NSC-UHPFRC composite. The $N1$ values were in the range of 12 blows to 37 blows; the majority of test samples in this group demonstrated a great ability to resist impact stress prior to the appearance of the first crack. At the failure stage, the minimum and maximum blows were 17 and 46 blows, respectively. The statistical indicators, mean, SD, and COV were 24, 6.49 blows, and 26.92%, respectively. Figures 9 and 10 present the distribution and probability plots of the impact strength data of the control NSC. The distribution graphs consisted of frequency histograms and superimposed normal curves. The test results at the two cracking stages nearly followed the normal distribution, as depicted in Figure 9a,b. The data were distributed near the mean value. The probability plots at the two cracking phases are shown in Figure 10a,b. It was observed that data points converged at the fitted line. The Kolmogorov–Smirnov (K–S) test was used to prove the normal distribution outcomes at a 0.05 significance level, and the results indicated that $N1$ had a p-value = 0.8926 and $N2$ had a p-value = 1. The results are consistent with normal distribution analysis.

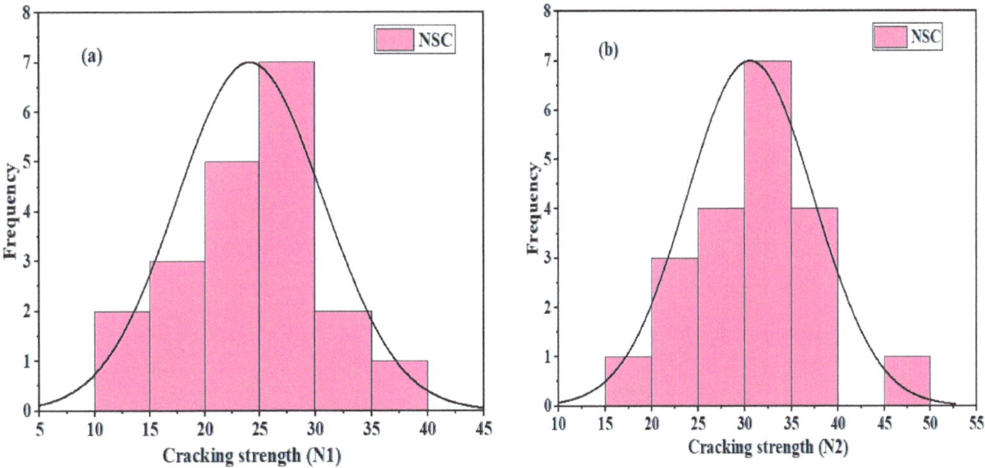

Figure 9. Distribution plots of impact test results for NSC: (**a**) $N1$ and (**b**) $N2$.

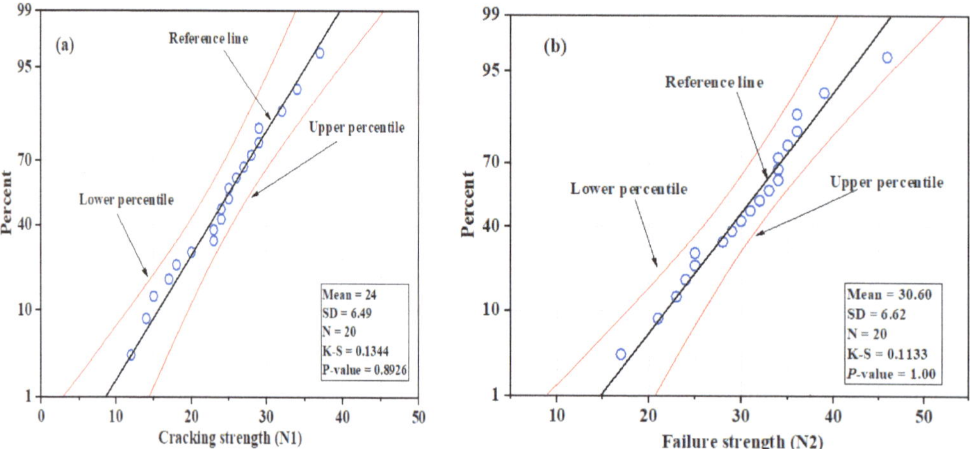

Figure 10. Normal probability plots of impact test results for NSC: (**a**) *N1* (**b**) *N2*.

3.3. NSC-UHPFRC-Nst

The impact test results for U-shaped NSC-UHPFRC at the first and fracture stages were in the range of 10–28 blows for N1 and 15–37 blows for N2, as summarized in Table 4, with mean and SD values of 21.6 blow and 5.52 blows, respectively. For *N1*, the mean and standard deviation at the fracture stage stood at 26.5 blows and 5.58 blows, respectively. The impact test results of this group showed a slight reduction in the coefficient of variation compared to the control specimens; COV values of 25.54% and 21.06% were obtained for N1 and *N2*, respectively. Most of the test specimens in this group showed a high ability to resist impact loads that initiate the first crack during testing, as presented in Table 4. The normal distribution and probability plots of NSC-UHPFR for *N1* and *N2* are presented in Figures 11 and 12. The plots show the impact strength data at the first and fracture stages follow a normal distribution, with distribution values located near the mean values in the overlaid curve, and the majority of data points concentrated along the fitted line, which is nearly normal. The K-S test result was at a 95% confidence level for *N1* and *N2*, and the *p*-values were = 1.00.

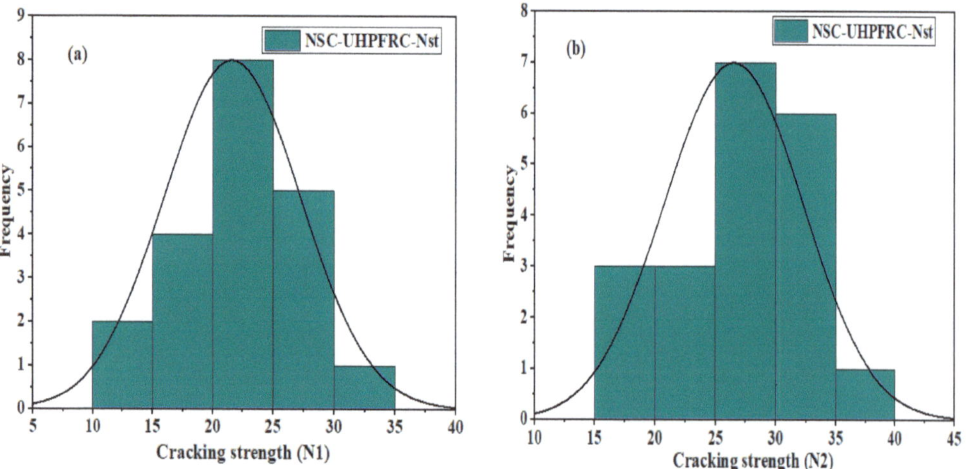

Figure 11. Distribution plots of NSC-UHPFRC-Nst impact test results: (**a**) *N1* and (**b**) *N2*.

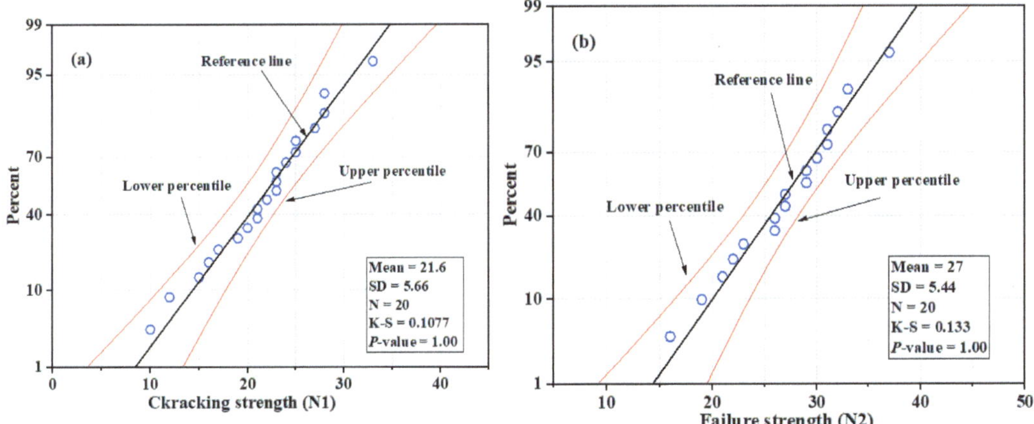

Figure 12. The probability plots of impact test results for NSC-UHPFRC-Nst: (**a**) *N1* (**b**) *N2*.

Table 4. Statistical indicator of the dataset.

	Parameters	Symbol	Units	Min	Max	Mean	STD	Kurtosis	Skewness
Input	Compressive strength of composite	*Cfc*	MPa	40.20	66.50	53.08	11.89	−1.99	0.017
	Flexural load	*P*	kN	1.31	3.21	2.58	0.75	−0.72	−1.06
	Density	*ρ*	kg/m	1946.1	2385.03	2068.62	73.24	3.21	1.24
	First crack strength	*N1*	blows	2	37	14.96	9.57	−1.04	0.29
Output	Failure strength	*N2*	blows	5	46	19.75	10.74	−1.03	0.15

3.4. NSC-UHPFRC-Sst

The U-shaped NSC-UHPFRC specimens treated with smooth surface treatment methods showed the lowest ability to withstand repeated impact loads. The first crack occurs easily when the impact is dropped. Most of the specimens exhibited a first crack with three blows; only one specimen withstood seven blows to induce first crack strength. Upon the addition of two impact blows, the specimens undergo complete failure, as noted in Table 4. The statistical indicators of mean, SD, and COV were 4.3 blows, 1.23 blows, and 35.3% at the initial crack phase, respectively. Figures 13 and 14 show the distribution and probability plots of the impact strength data of NSC-UHPFRC-Sst. The impact test at the first crack stage nearly follows a normal distribution, as depicted in Figure 13a. In contrast, the impact data at the fracture stage does not follow a normal distribution (see Figure 13b). The normal probability plots at the two cracking phases are shown in Figure 14a,b, which shows that data points are scattered around the fitting line. The Kolmogorov–Smirnov (K-S) test was used to prove the distribution results at a 0.05 significance level, and the results indicated that *N1* had a *p*-value = 0.0749 and *N2* had a *p*-value = 0.0651. The results are consistent with normal distribution analysis.

3.5. NSC-UHPFC-Gst

The impact test results of U-shaped NSC-UHPFRC-Gst samples were in the range of 5 blows–15 blows at the initial crack stage and in the range of 10 blows–22 blows at the failure stage, as summarized in Table 4, with mean and standard deviation values of 10.75 blows and 3.1 blows, respectively, at the first crack stage and 15.95 blows and 3.41 blows, respectively, at the fracture stage. The *N1* and *N2* values in this group had a coefficient of variation of 27.89% and 20.84%, respectively. Grooving surface treatment showed better results compared to the smooth surface technique. However, natural fracture demonstrated better bonding strength at resisting impact compared with the grooving surface adopted in this study. Figures 15 and 16 indicate normal distributions; probability was at a 95%

confidence level for both $N1$ and $N2$, and the p-values of $N1$ and $N2$ for NSC-UHPFR-Gst were plotted. The plots show the impact strength data at the first stage, and the fracture stage follows a normal distribution, with distribution values located near the mean point in the overlaid curve, and many data points concentrated at the fitted line, which is nearly normal. The K-S test results for $N1$ and $N2$ at the 95% confidence level had p-values = 0.999.

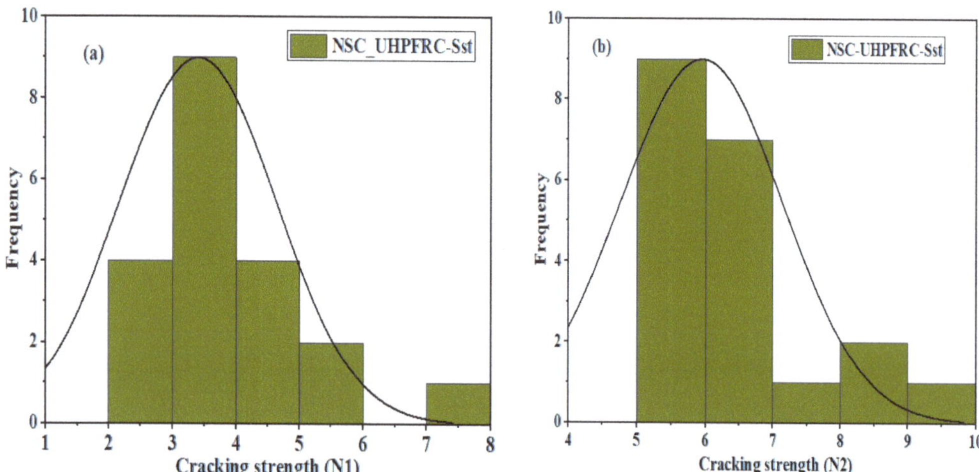

Figure 13. Distribution plots of NSC-UHPFRC-Sst impact test results: (**a**) $N1$ and (**b**) $N2$.

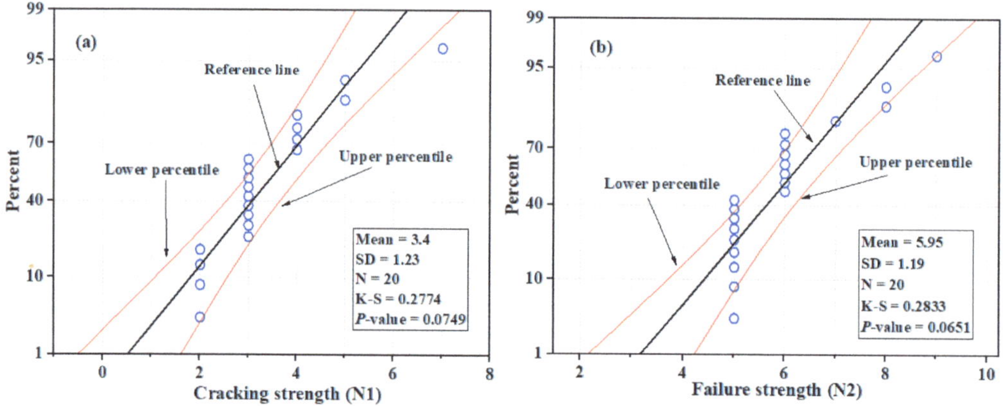

Figure 14. Probability plots of impact test results for NSC-UHPFRC-Sst: (**a**) $N1$ (**b**) $N2$.

3.6. Results of the Ensemble Machine Learning Models

This paper utilized the Python programming language to develop two ensemble machine-learning models. Each developed model was validated via a 10-fold cross-validation procedure [8,40]. XGBoost and CatBoost ensemble models were trained and evaluated using a dataset. The dataset included the first fracture strength ($N1$) compressive strength of composite (Cfc), Flexural load (P), and density (ρ) as the input variables. Failure strength ($N2$) was the target variable in the dataset.

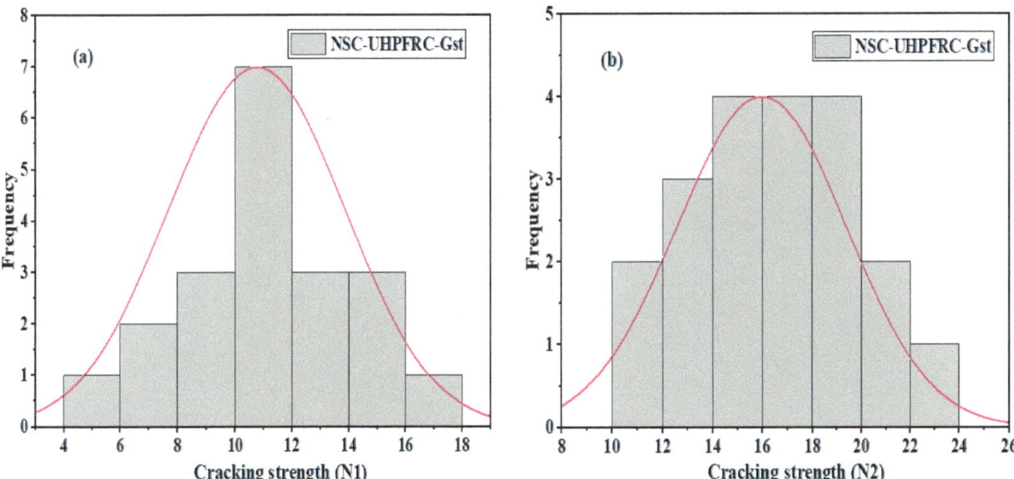

Figure 15. Distribution plots of NSC-UHPFRC-Gst impact test results: (**a**) *N1* and (**b**) *N2*.

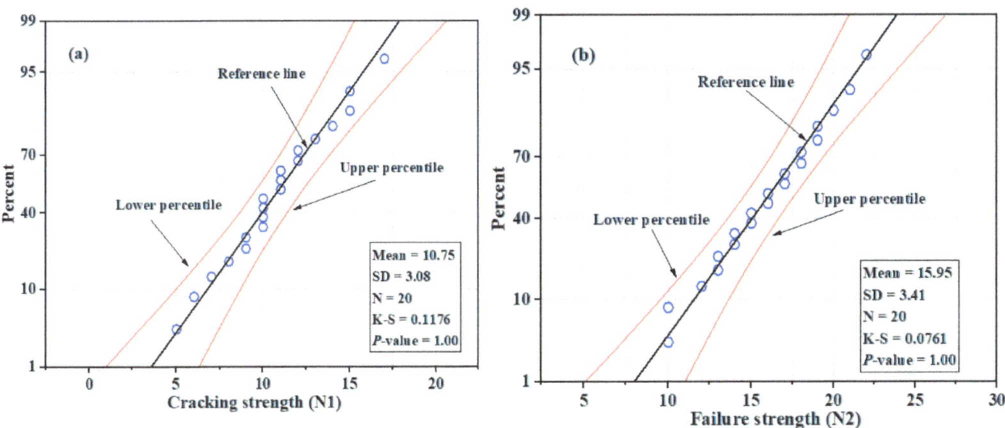

Figure 16. Normal probability plots of impact test results for NSC-UHPFRC-Gst: (**a**) *N1* (**b**) *N2*.

3.6.1. Sensitivity Analysis

The efficacy of ML techniques depends mostly on the capability of the input variables. Numerous input parameter involvement can lead to overfitting and complicate the generated model [41]. On the other hand, more input parameters can lead to accurate models. Therefore, as illustrated in Figure 17, sensitivity analysis employing Pearson correlation was utilized in this research to investigate the most credible input features for forecasting *N2*. Table 4 summarizes the statistical indicators of the dataset. Sensitivity analysis of the dataset showed that all the input variables are positively correlated with *N2*. However, *N1* is the most relevant variable, followed by the Flexural load, with a correlation coefficient value of 0.77. Moreover, compressive strength of the composite and density are correlated with *N2*, with correlation coefficient values of 0.32 and 0.31, respectively. The frequency distribution of the datasets is shown in Figure 18. From Figure 18, some of the variables follow the normal distribution, for example, density and *N2*, while some variables do not follow the normal distribution.

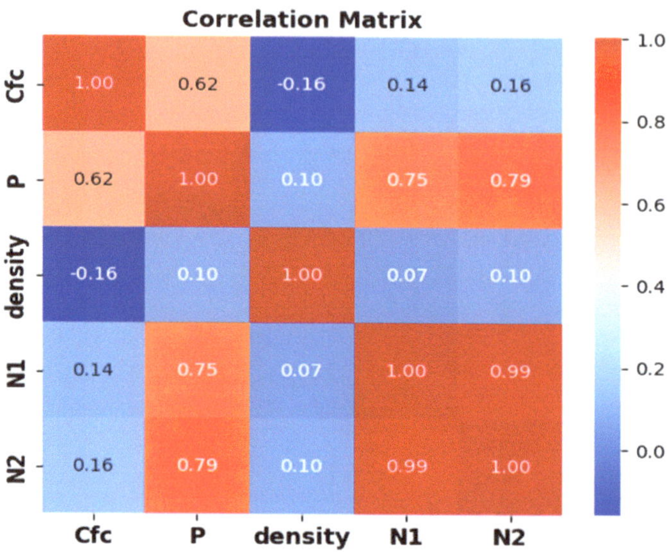

Figure 17. Pearson correlation matrix of the dataset.

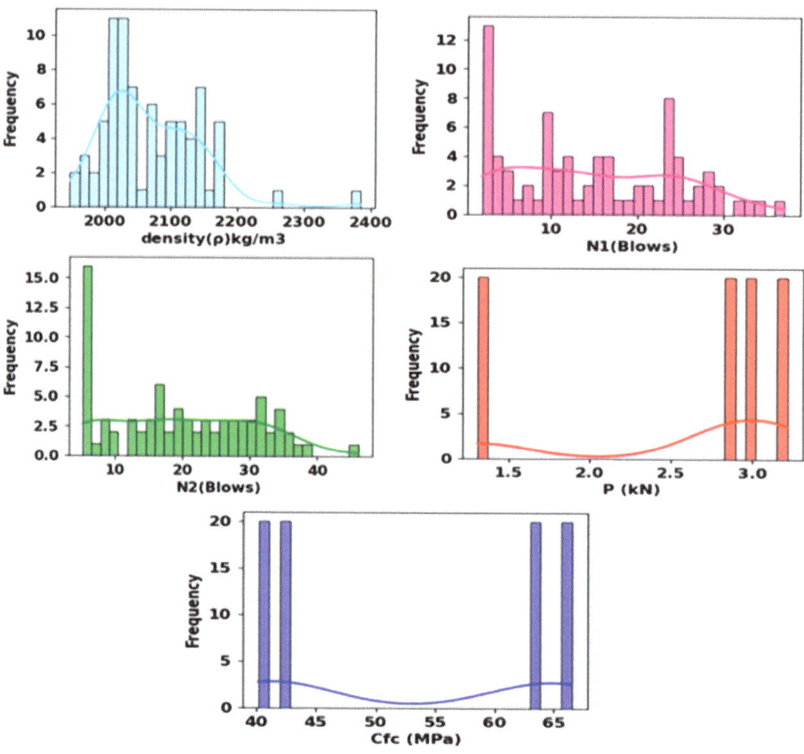

Figure 18. Relative distribution of the dataset parameters.

3.6.2. XGBoost Model Results

The prediction skills of the XGBoost model yielded a good relationship between the forecasted $N2$ and observed $N2$ values at both the training and testing stages. The model achieved an R^2 value of almost 0.999 at the training stage and an R^2 value of 0.9644 at the testing phase. The model revealed a lower value of performance indicators, as presented in Table 5. The model exhibited MSE = 0.000035, RMSE = 0.0019, and MAE = 0.0014 blows at the training phase. XGBoost outperformed all other models. Additionally, high prediction accuracy was achieved with Catboost, and reasonable prediction accuracy was obtained at the testing stage, showing better performance than SVM and GLM models at the training stage. The performance of the XGBoost model is shown in Figure 19a.

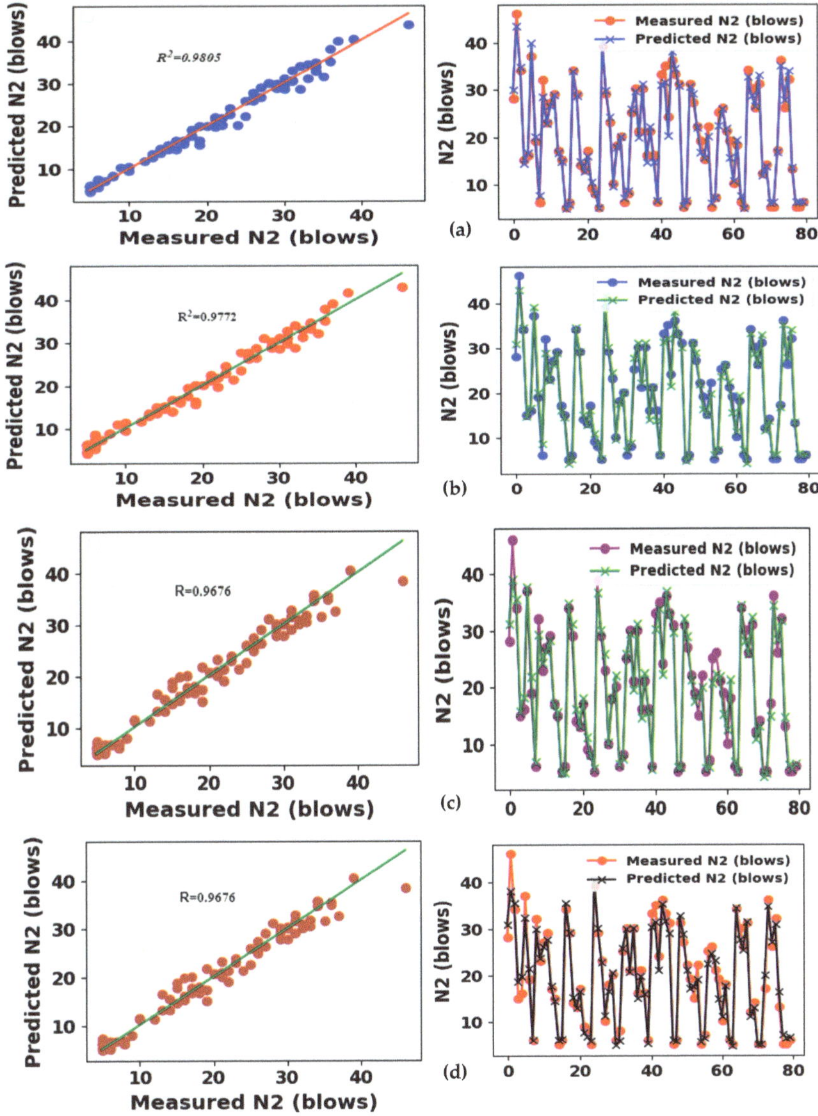

Figure 19. Scatter and line plots of the relationship between the predicted $N2$ and measured $N2$ values based on the (**a**) XGBoost, (**b**) Catboost (**c**), SVM and (**d**) GLM models.

Table 5. Model Performance.

Model	Training				Testing			
	MSE	RMSE	MAE	R^2	MSE	RMSE	MAE	R^2
XGBoost	0000035	0.0019	0.0014	1.0000	0.0000	0.0019	0.0014	0.9644
CatBoost	0.0638	0.2512	0.1950	0.9994	3.6676	1.6841	1.4128	0.9676
SVM	1.8663	1.3640	1.0410	0.9836	2.1014	1.4104	1.13836	0.9772
GLM	1.8784	1.3696	1.0395	0.9835	1.8784	1.3000	1.0395	0.9805

3.6.3. CatBoost Model Results

As obtained by the XGBoost model, CatBoost also predicts $N2$ with good accuracy. The R^2 of the CatBoost model at the training phase is 0.994, which is remarkably close to that of the XGBoost model. Additionally, Catboost outperformed the other models at the testing phase, with an R^2 value of 0.9671, as shown in Figure 19b, which is slightly lower than the accuracy obtained with the SVM and GLM models. MSE, RMSE, and MAE values of 0.0368, 0.2512, and 0.1950 blows were found for the CatBoost model, respectively, at the testing stage, indicating lower performance compared to XGBoost in terms of the metrics used. Furthermore, the measures of performance applied to the ensemble models had lower errors compared with the two ensemble models, as outlined in Table 5.

3.6.4. SVR and GLM

The SVM also performs very well at predicting the failure strength ($N2$), with an R^2 value of 0.9772, as presented in Figure 19c. This shows a higher performance accuracy than XGBoost and CatBoost, but slightly lower performance accuracy compared to a general linear model. The MSE, RMSE, and MEA values of SVM at the testing stage were 2.1014, 1.4104, and 1.1383, respectively. Assessment of the failure strength ($N2$) by the GLM model yields the highest coefficient of determination at the testing stage, giving an R^2 value of 0.9805, in contrast to XGBoost, CatBoost, and SVM as indicated in Table 5 and depicted in Figure 19d. MSE, RMSE, and MAE values of 1.88784, 1.3000, and 1.0395, respectively, were obtained at the testing phase using the GLM model, which indicates higher performance compared with CatBoost and SVM.

The four (4) developed models, Catboost, XGboost, SVR, and GLM, were also analyzed based on the two-dimensional space diagram called the Taylor diagram (Figure 20), which graphically shows the measured and predicted values [42]. The Taylor diagram was recognized for its accurate judgment [43]. Two measurements, SD and correlation (R), were combined to build the Taylor diagram [44]. The primary purpose of this diagram is to evaluate various performance metrics in one combination and statistically compute the degree of similarity between the observed and anticipated $N2$ values. It can be noted that in both the training and testing phases (Figure 20a,b), the best performance results for the XGboost models, with an R-value = 0.999, were obtained at the training stage. Based on the data, the specified indicator represents the level of prediction accuracy for XGboost. As a result, XGboost outperformed other models because the observed locations were closer to the computed values. This is further supported by the high value of SD that was attributed to the XGboost model. In general, overestimation occurs when the SD of computed values exceeds the SD of measured values.

In addition, the model performance was evaluated using the percentage relative error, shown in Figure 21. The relative error achieved by each model was compared at the two modeling stages, as depicted in Figure 21. From Figure 21a, Catboost and XGboost proved to be the best models, with the least and highest relative error distributions for forecasting $N2$. The Catboost and XGboost models' first and third quartile (Q1 and Q3) values appeared close to one another, while GLM appeared to be the best-performing model at the testing stage.

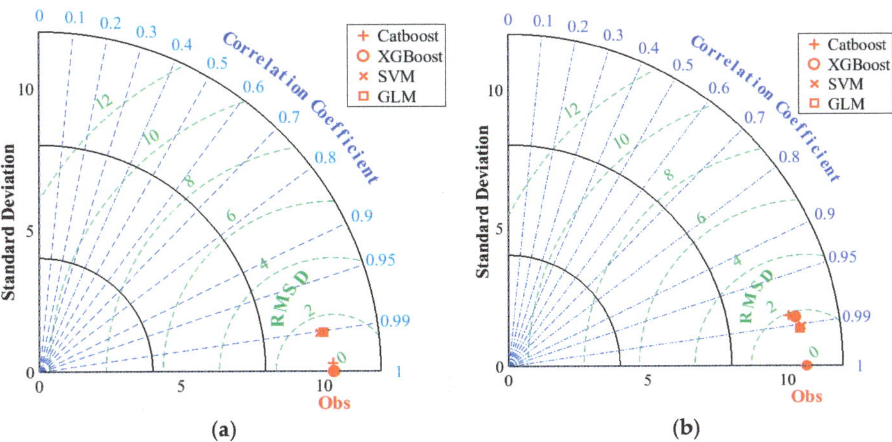

Figure 20. Taylor diagrams displaying model performance: (**a**) training phase and (**b**) testing phase.

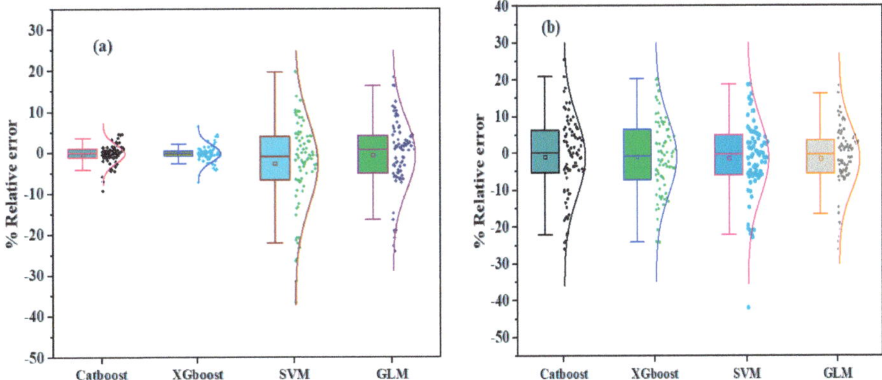

Figure 21. Boxplots displaying relative error distributions: (**a**) training and (**b**) testing stages.

4. Conclusions

The study investigated the bonding behavior of composite U-shaped NSC-UHPFRC samples using multiple drop-weight impact testing methods. The interface between the NSC substrate and the UHPFRC layer was treated with three surface treatment systems: grooving, natural fracture, and smoothing techniques. Ensemble machine learning algorithms (comprising XGBoost and CatBoost), SVM, and GLM, were employed to train and test the simulation dataset to estimate the impact strength ($N2$) of the U-shaped NSC-UHPFRC specimen. The findings of the research are summarized, and the conclusions are stated below:

(1)　The impact test result indicated that surface treatment plays a significant role in ensuring sufficient bond strength at the interface of NSC-UHPFRC composites, and the bond behavior between the NSC substrate under natural fracture and the UHPFRC layer can provide sufficient bond strength at the interface, resulting in a monolithic structure that can withstand dynamic loads under repeated drop-weight impact stress.

(2)　The reference NSC specimen requires a high number of drops to resist impact loads before initial crack ($N1$) occurrence, with the average number of drops equal to 24 blows compared to the NSC-UHPFRC composite samples. Remarkable reductions in impact strength properties were observed in all the composite U-shaped NSC-UHPFRC samples.

(3)   The inclusion of steel fibers in the UHPFRC layer improved the composite U-shaped ductility, which transformed the composite specimens to a more ductile state and enhanced the impact strength of the NSC-UHPFRC sample. The DI values of the NSC-UHPFRC for each testing condition are less than unity. The COV of the impact data obtained in this work is lower than that found in several past studies that used the drop-weight impact testing approach ACI 544-2R.

(4)   The two ensemble ML approaches correctly estimated the impact strength of the NSC-UHPFRC composite. The XGBoost ensemble model gave coefficient of determination (R^2) values of approximately 0.999 and 0.964 at the training and testing stages. Similarly, GLM outperformed other models in the testing phase, with an R^2 value of 0.9805. The performance matrices were proven using the Taylor diagram and Boxplots.

Author Contributions: S.I.H. and I.H.H.: conceptualization, methodology, investigation, writing—original draft, writing—review and editing; Y.E.I.: writing—original draft, writing—review and editing, and funding acquisition. A.A.-s. and H.Z.: Data curation, visualization, writing—original draft, writing—review and editing. All authors have read and agreed to the published version of the manuscript.

Funding: The authors greatly acknowledge financial support for this research from the Structure and Material (S&M) Lab of Prince Sultan University, which funded the APC. The authors greatly acknowledge financial support for this research from the Natural Science Foundation of China (No. 51708314) and the Structures and Materials Laboratory (S&M Lab) of the College of Engineering, Prince Sultan University, Riyadh, Saudi Arabia, for funding the article process fees.

Institutional Review Board Statement: Not applicable.

Informed Consent Statement: Not applicable.

Data Availability Statement: Data are contained within the article.

Conflicts of Interest: The authors declare no conflicts of interest.

References

1. Semendary, A.A.; Svecova, D. Factors affecting bond between precast concrete and cast in place ultra high performance concrete (UHPC). *Eng. Struct.* **2020**, *216*, 110746. [CrossRef]
2. Beushausen, H.; Höhlig, B.; Talotti, M. The influence of substrate moisture preparation on bond strength of concrete overlays and the microstructure of the OTZ. *Cem. Concr. Res.* **2017**, *92*, 84–91. [CrossRef]
3. Feng, S.; Xiao, H.; Geng, J. Bond strength between concrete substrate and repair mortar: Effect of fibre stiffness and substrate surface roughness. *Cem. Concr. Compos.* **2020**, *114*, 103746. [CrossRef]
4. Mansour, W.; Sakr, M.A.; Seleemah, A.A.; Tayeh, B.A.; Khalifa, T.M. Bond behavior between concrete and prefabricated ultra high-performance fiber-reinforced concrete (UHPFRC) plates. *Struct. Eng. Mech.* **2022**, *81*, 305.
5. Yu, J.; Zhang, B.; Chen, W.; Liu, H.; Li, H. Multi-scale study on interfacial bond failure between normal concrete (NC) and ultra-high performance concrete (UHPC). *J. Build. Eng.* **2022**, *57*, 104808. [CrossRef]
6. Yang, Y.; Xu, C.; Yang, J.; Wang, K. Experimental study on flexural behavior of precast hybrid UHPC-NSC beams. *J. Build. Eng.* **2023**, *70*, 106354. [CrossRef]
7. Jongvivatsakul, P.; Thongchom, C.; Mathuros, A.; Prasertsri, T.; Adamu, M.; Orasutthikul, S.; Lenwari, A.; Charainpanitkul, T. Enhancing bonding behavior between carbon fiber-reinforced polymer plates and concrete using carbon nanotube reinforced epoxy composites. *Case Stud. Constr. Mater.* **2022**, *17*, e01407. [CrossRef]
8. Haruna, S.I.; Zhu, H.; Jiang, W.; Shao, J. Evaluation of impact resistance properties of polyurethane-based polymer concrete for the repair of runway subjected to repeated drop-weight impact test. *Constr. Build. Mater.* **2021**, *309*, 125152. [CrossRef]
9. Haruna, S.I.; Zhu, H.; Ibrahim, Y.E.; Shao, J.; Adamu, M.; Farouk, A.I.B. Experimental and Statistical Analysis of U-Shaped Polyurethane-Based Polymer Concrete under Static and Impact Loads as a Repair Material. *Buildings* **2022**, *12*, 1986. [CrossRef]
10. Haruna, S.I.; Zhu, H.; Ibrahim, Y.E.; Shao, J.; Adamu, M.; Ahmed, O.S. Impact resistance and flexural behavior of U-shaped concrete specimen retrofitted with polyurethane grout. Case Stud. *Constr. Mater.* **2023**, *19*, e02547.
11. Jia, P.C.; Wu, H.; Wang, R.; Fang, Q. Dynamic responses of reinforced ultra-high performance concrete members under low-velocity lateral impact. *Int. J. Impact Eng.* **2021**, *150*, 103818. [CrossRef]
12. Zhai, Y.X.; Wu, H.; Fang, Q. Impact resistance of armor steel/ceramic/UHPC layered composite targets against 30CrMnSiNi2A steel projectiles. *Int. J. Impact Eng.* **2021**, *154*, 103888. [CrossRef]

13. Zhang, F.; Shedbale, A.S.; Zhong, R.; Poh, L.H.; Zhang, M.-H. Ultra-high performance concrete subjected to high-velocity projectile impact: Implementation of K&C model with consideration of failure surfaces and dynamic increase factors. *Int. J. Impact Eng.* **2021**, *155*, 103907.

14. Huang, L.; Su, L.; Xie, J.; Lu, Z.; Li, P.; Hu, R.; Yang, S. Dynamic splitting behaviour of ultra-high-performance concrete confined with carbon-fibre-reinforced polymer. *Compos. Struct.* **2022**, *284*, 115155. [CrossRef]

15. Wu, Z.; Shi, C.; Khayat, K.H.; Xie, L. Effect of SCM and nano-particles on static and dynamic mechanical properties of UHPC. *Constr. Build. Mater.* **2018**, *182*, 118–125. [CrossRef]

16. Yalin, L.; Ya, W. Drop-Weight Impact Resistance of Ultrahigh-Performance Concrete and the Corresponding Statistical Analysis. *J. Mater. Civ. Eng.* **2022**, *34*, 4021409.

17. Yu, R.; Van Beers, L.; Spiesz, P.; Brouwers, H.J.H. Impact resistance of a sustainable Ultra-High Performance Fibre Reinforced Concrete (UHPFRC) under pendulum impact loadings. *Constr. Build. Mater.* **2016**, *107*, 203–215. [CrossRef]

18. Almustafa, M.K.; Nehdi, M.L. Machine learning model for predicting structural response of RC columns subjected to blast loading. *Int. J. Impact Eng.* **2022**, *162*, 104145. [CrossRef]

19. Cao, Y.; Zandi, Y.; Rahimi, A.; Petković, D.; Denić, N.; Stojanović, J.; Spasić, B.; Vujović, V.; Khadimallah, M.A.; Assilzadeh, H. Evaluation and monitoring of impact resistance of fiber reinforced concrete by adaptive neuro fuzzy algorithm. *Structures.* **2021**, *34*, 3750–3756. [CrossRef]

20. Khosravani, M.R.; Nasiri, S.; Anders, D.; Weinberg, K. Prediction of dynamic properties of ultra-high performance concrete by an artificial intelligence approach. *Adv. Eng. Softw.* **2019**, *127*, 51–58. [CrossRef]

21. Shao, R.; Wu, C.; Su, Y.; Liu, Z.; Liu, J.; Chen, G.; Xu, S. Experimental and numerical investigations of penetration resistance of ultra-high strength concrete protected with ceramic balls subjected to projectile impact. *Ceram. Int.* **2019**, *45*, 7961–7975. [CrossRef]

22. GB175-2007; Common Portland Cement. Chinese National Standard: Beijing, China, 2007.

23. JTG55-2011; Specification for Mix Proportion Design of Ordinary Concrete. Chinese National Standard: Beijing, China, 2011.

24. Farouk, A.I.B.; Harunaa, S.I. Evaluation of Bond Strength between Ultra-High-Performance Concrete and Normal Strength Concrete: An Overview. *J. Kejuruter.* **2020**, *32*, 41–51. [CrossRef] [PubMed]

25. ACI 544.2R-89; Measurement of Properties of Fiber Reinforced Concrete. American Concrete Institute: Farmington Hills, MI, USA, 1989; pp. 433–439.

26. Al-shawafi, A.; Zhu, H.; Bo, Z.; Haruna, S.I.; Ibrahim, Y.E.; Farouk, A.I.B.; Laqsum, S.A.; Shao, J. Bond behavior between normal concrete and UHPC and PUC layers subjected to different loading conditions coupled with fracture analysis technique. *J. Build. Eng.* **2024**, *86*, 108880. [CrossRef]

27. Malami, S.I.; Musa, A.A.; Haruna, S.I.; Aliyu, U.U.; Usman, A.G.; Abdurrahman, M.I.; Bashir, A.; Abba, S.I. Implementation of soft-computing models for prediction of flexural strength of pervious concrete hybridized with rice husk ash and calcium carbide waste. *Model. Earth Syst. Environ.* **2022**, *8*, 1933–1947. [CrossRef]

28. Qian, L.; Chen, Z.; Huang, Y.; Stanford, R.J. Employing categorical boosting (CatBoost) and meta-heuristic algorithms for predicting the urban gas consumption. *Urban Clim.* **2023**, *51*, 101647. [CrossRef]

29. Younsi, S.; Dabiri, H.; Marini, R.; Mazzanti, P.; Mugnozza, G.S.; Bozzano, F. Reconstructing missing InSAR data by the application of machine leaning-based prediction models: A case study of Rieti. *J. Civ. Struct. Health Monit.* **2024**, *14*, 143–161. [CrossRef]

30. Dabiri, H.; Clementi, J.; Marini, R.; Mugnozza, G.S.; Bozzano, F.; Mazzanti, P. Machine learning-based analysis of historical towers. *Eng. Struct.* **2024**, *304*, 117621. [CrossRef]

31. Lyu, F.; Fan, X.; Ding, F.; Chen, Z. Prediction of the axial compressive strength of circular concrete-filled steel tube columns using sine cosine algorithm-support vector regression. *Compos. Struct.* **2021**, *273*, 114282. [CrossRef]

32. Brereton, R.G.; Lloyd, G.R. Support vector machines for classification and regression. *Analyst* **2010**, *135*, 230–267. [CrossRef]

33. Nelder, J.A.; Wedderburn, R.W.M. Generalized linear models. *J. R. Stat. Soc. Ser. A Stat. Soc.* **1972**, *135*, 370–384. [CrossRef]

34. Cao, H. Experimental investigation on the static and impact behaviors of basalt fiber-reinforced concrete. *Open Civ. Eng. J.* **2017**, *11*, 14–21. [CrossRef]

35. Rai, B.; Singh, N.K. Statistical and experimental study to evaluate the variability and reliability of impact strength of steel-polypropylene hybrid fiber reinforced concrete. *J. Build. Eng.* **2021**, *44*, 102937. [CrossRef]

36. Prasad, N.; Murali, G. Exploring the impact performance of functionally-graded preplaced aggregate concrete incorporating steel and polypropylene fibres. *J. Build. Eng.* **2021**, *35*, 102077. [CrossRef]

37. Badr, A.; Ashour, A.F.; Platten, A.K. Statistical variations in impact resistance of polypropylene fibre-reinforced concrete. *Int. J. Impact Eng.* **2006**, *32*, 1907–1920. [CrossRef]

38. Song, P.S.; Wu, J.C.; Hwang, S.; Sheu, B.C. Statistical analysis of impact strength and strength reliability of steel–polypropylene hybrid fiber-reinforced concrete. *Constr. Build. Mater.* **2005**, *19*, 1–9. [CrossRef]

39. Haruna, S.I.; Zhu, H.; Shao, J. Experimental study, modeling, and reliability analysis of impact resistance of micro steel fiber-reinforced concrete modified with nano silica. *Struct. Concr.* **2022**, *23*, 1659–1674. [CrossRef]

40. Farouk, A.I.B.; Jinsong, Z. Prediction of Interface Bond Strength between Ultra-High-Performance Concrete (UHPC) and Normal Strength Concrete (NSC) Using a Machine Learning Approach. *Arab. J. Sci. Eng.* **2022**, *47*, 5337–5363. [CrossRef]

41. Ahmed, A.A.; Pradhan, B. Vehicular traffic noise prediction and propagation modelling using neural networks and geospatial information system. *Environ. Monit. Assess.* **2019**, *191*, 190. [CrossRef] [PubMed]

42. Yaseen, Z.M.; Ebtehaj, I.; Kim, S.; Sanikhani, H.; Asadi, H.; Ghareb, M.I.; Bonakdari, H.; Mohtar, W.H.W.; Al-Ansari, N.; Shahid, S. Novel Hybrid Data-Intelligence Model for Forecasting Monthly Rainfall with Uncertainty Analysis. *Water* **2019**, *11*, 502. [CrossRef]
43. Zhu, S.; Heddam, S.; Wu, S.; Dai, J.; Jia, B. Extreme learning machine-based prediction of daily water temperature for rivers. *Environ. Earth Sci.* **2019**, *78*, 202. [CrossRef]
44. Taylor, K.E. Summarizing multiple aspects of model performance in a single diagram. *J. Geophys. Res.* **2001**, *107*, 7183–7192. [CrossRef]

Article

Alternative Method for Determination of Vibroacoustic Material Parameters for Building Applications

Krzysztof Nering [1,*] and **Konrad Nering** [2]

1. Faculty of Civil Engineering, Cracow University of Technology, 31-155 Cracow, Poland
2. Faculty of Mechanical Engineering, Cracow University of Technology, 31-155 Cracow, Poland; konrad.nering@pk.edu.pl
* Correspondence: krzysztof.nering@pk.edu.pl

Abstract: The development of urbanization and the resulting expansion of residential and transport infrastructures pose new challenges related to ensuring comfort for city dwellers. The emission of transport vibrations and household noise reduces the quality of life in the city. To counteract this unfavorable phenomenon, vibration isolation is widely used to reduce the propagation of vibrations and noise. A proper selection of vibration isolation is necessary to ensure comfort. This selection can be made based on a deep understanding of the material parameters of the vibration isolation used. This mainly includes dynamic stiffness and damping. This article presents a comparison of the method for testing dynamic stiffness and damping using a single degree of freedom (SDOF) system and the method using image processing, which involves tracking the movement of a free-falling steel ball onto a sample of the tested material. Rubber granules, rubber granules with rubber fibers, and rebound polyurethanes were selected for testing. Strong correlations were found between the relative indentation and dynamic stiffness (at 10–$60 \ \mathrm{MN/m^3}$) and the relative rebound and damping (for 6–12%). Additionally, a very strong relationship was determined between the density and fraction of the critical damping factor/dynamic stiffness. The relative indentation and relative rebound measurement methods can be used as an alternative method to measure the dynamic stiffness and critical damping factor, respectively.

Keywords: acoustic comfort; vibrational comfort; material properties; damping; dynamic stiffness; polyurethane; rubber granulate; image processing; rebound; indentation

Citation: Nering, K.; Nering, K. Alternative Method for Determination of Vibroacoustic Material Parameters for Building Applications. *Materials* **2024**, *17*, 3042. https://doi.org/10.3390/ma17123042

Academic Editors: Honglei Chang, Li Ai, Feng Guo and Xinxiang Zhang

Received: 2 May 2024
Revised: 10 June 2024
Accepted: 17 June 2024
Published: 20 June 2024

1. Introduction

More and more people live in cities, and this trend only seems likely to continue [1–6]. The growth of the urban population causes the development of urbanization and, consequently, the transport infrastructure, which must meet increasingly greater challenges [7–11]. Urban planners make compromises related to the proximity of residential buildings to the railway and road infrastructure. This translates into shortening the travel time of city residents, which is undoubtedly a desirable phenomenon [12–14]. However, the price for such a development is the increased presence of bothersome stimuli, such as noise and vibrations, especially from the railway infrastructure [15–22].

The development of urbanization also imposes the need to increase population density, which is reflected not only in the way buildings are located, but also in the increase in the number of inhabitants in a single building [23–27]. This is a necessary response to the market situation [28–30], but the consequence is also a reduction in the comfort of residents by increasing the noise from neighbors to which residents are exposed [28,31–33].

The expansion of transport infrastructure and housing construction, undoubtedly beneficial to national economies, reduces the quality of life. Therefore, it is necessary to consciously design the infrastructure, taking into account the nuisance of its use. A reduction in harmful stimuli such as noise and vibration is required to ensure the health and well-being of urban inhabitants. The answer to the above challenges is the introduction of appropriate directives, regulations [34,35], and standards [36–42] that constitute the means and the goal to be achieved as well as the methodology for achieving this goal. The constantly developed approach to assessing comfort requires a good understanding of how humans perceive bothersome stimuli [21,43–47]. Ways to ensure comfort require a lot of information about the technology and materials used [48,49].

There are various ways to reduce the emission of harmful stimuli [50–55]. It is worth mentioning the use of vibration isolation on railway tracks, which allows for the reduction in the propagation of vibrations generated during the passage of a train [56–58]. This has a positive impact not only on buildings located in the vibration influence zone [59–63], but also on residents who are exposed to a smaller dose of annoying vibrations [64–67].

In residential and office buildings, i.e., places intended for people to stay over long periods at a time, noise protection is used in the form of floating floors or raised flooring [68–72]. Partitions are selected appropriately so that they increase acoustic insulation. This allows for the reduction in exposure to airborne and impact noise [73–76].

Also, with the increasing popularity of mechanical ventilation, which is dictated by energy efficiency [77–79], it is worth mentioning the use of passive vibration isolation of technical devices and duct silencers in order to reduce noise emissions related to the operation of ventilation system devices [80–84].

Vibration isolation also plays an important role in controlling seismic actions on the building. A proper assessment of vibration isolation parameters such as stiffness and damping is key to predict a building's response to seismic actions [85–87]. An appropriate design of the structure allows for the control of resonance frequencies in the context of seismic forces and appropriate control over the internal forces in the structure [88–90].

The dominant material in passive vibration isolation used in various construction sectors is rubber [91–95]. This material has very good features related to stiffness, dimensional stability, and damping. Materials derived from waste are also used, such as rubber granules [96–98] or polyurethanes [99–102]. The fact that polyurethanes can be recycled materials increases their potential in terms of sustainable development. In residential construction, elastified polystyrene and mineral wool dominate as vibration-insulating materials for floors [103–107], but there is a potential to use rebound polyurethanes because they have appropriate stiffness parameters.

The most important material parameters from the point of view of vibration isolation are dynamic stiffness and damping [108–115]. These parameters allow for the adequate design of a passive vibration isolation system. This allows the designer to properly assess if a given material is suitable for the added solution. For residential construction, the material assessment method is determined according to the ISO 9052-1 standard [48], where the tests allow for the estimation of the dynamic stiffness under small stresses (2 kPa). Additionally, the second parameter, namely, the damping of the material, can also be assessed; this will have a significant impact when designing an experiment, considering the impact of vibrations on people [100,115,116].

The aim of this article is to validate an alternative method for estimating the value of dynamic stiffness at low stresses [48] and damping [100] using image analysis during a free ball drop. This study is intended to allow for a faster assessment of these parameters. The usefulness of the alternative method can be indicated especially during preliminary research—for the purpose of the preliminary selection of materials or estimation of the above-mentioned parameters for research purposes.

2. Measurement Methodology and Setup

The selected materials for the research were vibration-isolating materials: rebound polyurethane, rubber granules, and rubber granules with rubber fibers. The dimensions of the samples were 200 mm × 200 mm × 50 mm. The compilation of material parameters is presented in Table 1.

Table 1. List of materials used for research.

Material Name	Sample ID	Density (kg/m^3)	Density Averaged (kg/m^3)
rubber granulate (S1000)	S1000_01	1008.0	1017.25
	S1000_02	1014.0	
	S1000_03	1023.0	
	S1000_04	1024.0	
rubber granulate (S850)	S850_01	840.5	865.25
	S850_02	848.0	
	S850_03	870.5	
	S850_04	902.0	
rubber granulate with rubber fibers (F570)	F570_01	606.5	622.5
	F570_02	621.5	
	F570_03	626.0	
	F570_04	636.0	
rubber granulate with rubber fibers (F700)	F700_01	697.5	719.5
	F700_02	709.5	
	F700_03	723.5	
	F700_04	747.5	
rebound polyurethane (P150)	P150_01	142.0	146.75
	P150_02	144.0	
	P150_03	150.0	
	P150_04	151.0	
rebound polyurethane (P250)	P250_01	243.0	257.25
	P250_02	254.0	
	P250_03	262.0	
	P250_04	270.0	

The samples tested in this article are presented in Figure 1.

Figure 1. Samples used for research in this article. Tested materials are rubber granulate (S), rubber granulate with rubber fibers (F), and rebound polyurethane (P).

The choice of materials was dictated by several factors. First of all, the selected materials were to be used as passive vibration isolation. Another factor was the widest possible variation in the density of the materials. The fact that the materials had to differ in stiffness and compressibility was also taken into account.

2.1. Dynamic Stiffness and Critical Damping Factor Estimation with Reference Method

In this paper, measurements were made using a dynamic stiffness testing machine [48,100], which determined the dynamic stiffness of the material and the critical damping fraction.

This method involved determining the dynamic response of the system, which was modelled using a single degree of freedom (SDOF) system as shown in Figure 2. Based on the dynamic response, the resonant frequency and the critical damping fraction were determined.

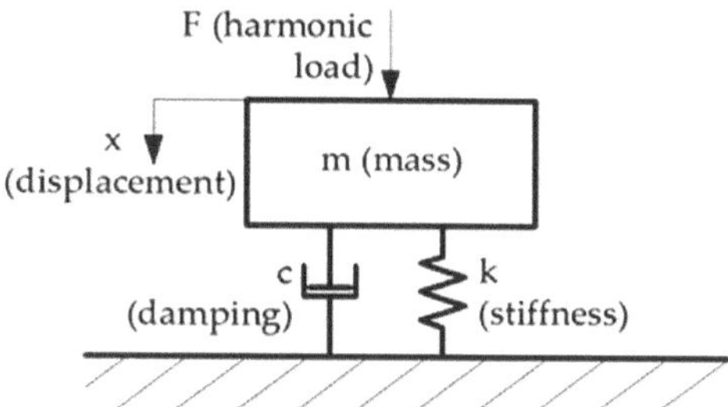

Figure 2. Physical model of the mass–damper–spring system with single degree of freedom (SDOF).

The resonance frequency was estimated by locating the maximum value in the vibration acceleration response spectrum from the frequency domain analysis of the response. Then, the resonant frequency was converted using the formula for the dynamic stiffness of the frame with Equation (1).

$$DS = 4\pi^2 m' f_r^2 \tag{1}$$

where DS is dynamic stiffness, m' is total mass per unit area used during the test, and f_r is resonant frequency

The critical damping fraction was estimated using the half-power method. The scheme for determining the critical damping fraction is shown in Figure 3.

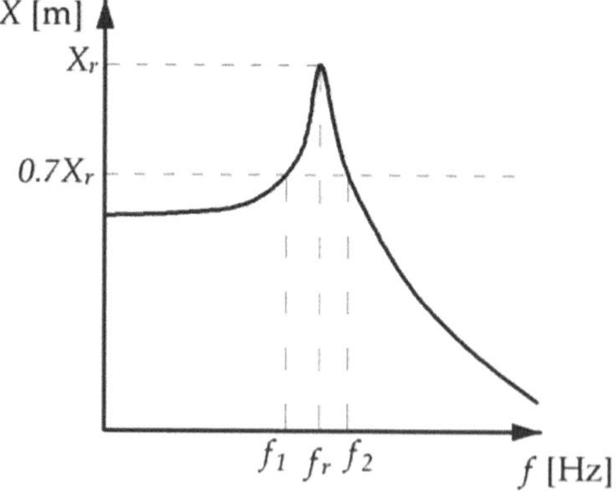

Figure 3. Schematic presentation of the half-power bandwidth method using the displacement spectrum. X_r—displacement amplitude, f_r—resonant frequency, and f_1 and f_2—correspond to frequencies to 0.7 value of resonance amplitude.

Once obtaining the values according to the above scheme, the damping relationships were described mathematically using Equation (2) [100,117,118].

$$\frac{f_2 - f_1}{f_r} = \frac{\delta}{\pi\sqrt{1 - \left(\frac{\delta}{2\pi}\right)^2}} \tag{2}$$

Equation (3) describes the relationship by which the critical damping fraction was determined [19,100].

$$\delta = \frac{2\pi D}{\sqrt{1 - D^2}} \tag{3}$$

The description of the parameters of the machine used for dynamic stiffness testing is presented in Table 2.

Table 2. Parameters of machine used for dynamic stiffness test.

Device Name/Manufacturer	Key Feature	Key Value of Parameters
Dynamic exciter—Brüel & Kjær (Virum, Denmark) Mini-shaker Type 4810	Provides sinusoidal force	Sine peak max 10 N Frequency range DC-18 kHz
IEPE accelerometer—MMF (Radebeul, Germany) KS78B.100	Measures acceleration of system response	Peak acceleration 60 g (~600 m/s^2) Linear frequency range (5% deviation) 0.6 Hz–14 kHz
Force sensor—Forsentek (Shenzhen, China) FSSM 50 N	Measures force applied to system	Max force 50 N Rated output 2.0 mV/V Hysteresis $\pm$ 0.1% R.O. (rated output)
Dynamic stiffness test bench	Measures resonant frequency of sample (200 mm × 200 mm × 50 mm) under load of 8 kg	Linear frequency range upper limit (5% deviation) 20–250 Hz—measured

The machine depicted in Figure 4 was utilized for conducting tests on vibroacoustic parameters. Primarily designed for assessing dynamic stiffness, this machine incorporates a dynamic exciter, which applies harmonic force to the load plate via a force sensor. The load plate, weighing 8 kg, is positioned atop the test sample with a pre-load static force ranging from 0.1 to 0.4 N. The system's response is measured using an IEPE (Integrated Electronics Piezo-Electric) accelerometer.

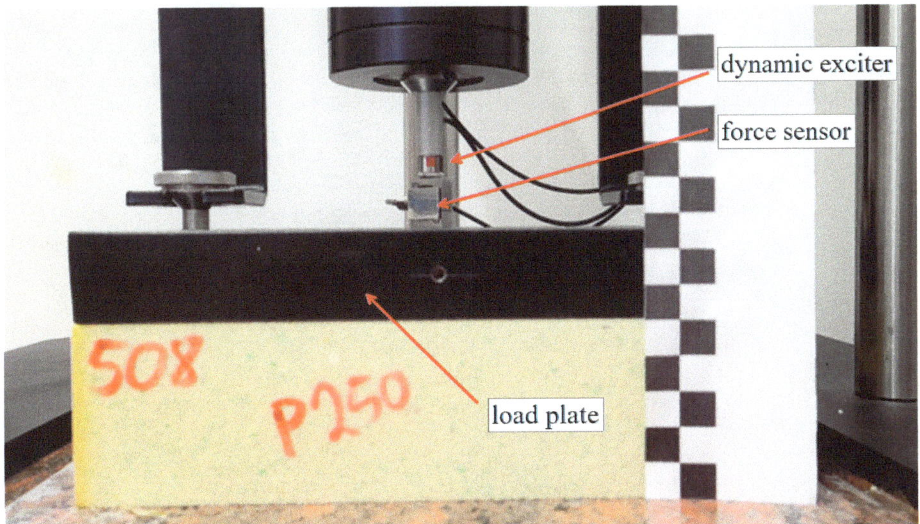

Figure 4. Dynamic stiffness test bench with loaded sample. One square in photograph is one centimeter.

The testing system was dynamically excited using a sinusoidal force generated by the exciter. The amplitude of the applied sinusoidal force was maintained at 0.4 N with a tolerance of ±0.005 N. The frequency range spanned from 1 Hz to 100 Hz. The frequency was incremented by 0.1 Hz every second during the measurement period. The system response was registered using an IEPE accelerometer positioned at the load plate. Exemplary result is shown in Figure 5.

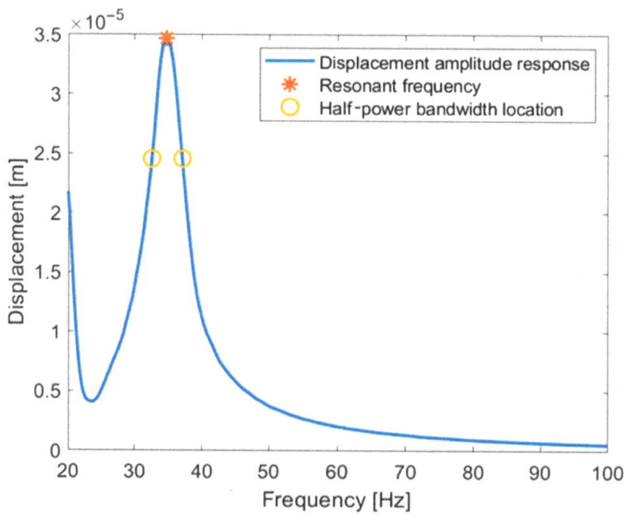

Figure 5. Pseudo-displacement response spectrum of sample P150 (288 kg/m^3) with resonant frequency of 35.2 Hz and critical damping factor 0.0645.

2.2. Alteranative Method

The image processing method was used as an alternative method proposed in this paper. This method involves tracking the movement of a steel ball in free fall from a given height and observing its behavior during and after rebound from the tested material. This method determines local extremes in the time course of the ball's position.

A high-speed camera was used to record the position of the ball. In order to identify local maxima, the optical axis of the camera lens was set halfway between the starting (drop) point and the upper surface of the sample. To record local minima, the optical axis of the camera lens was set in the plane of the upper surface of the sample.

In the research using this method, an Olympus (Tokyo, Japan) I-Speed high-speed camera with a frame rate of 1000 fps was used. An electromagnet released by a toggle switch was used to free drop the steel ball. In order to ensure good contrast to the falling ball, the scene was illuminated using a high-performance nonflickering LED reflector. To standardize the lighting, a diffuser made of white matt plexiglass was used. The ball was dropped into the center of the sample surface in order to uniform the stress distribution in the sample. The measurement setup is shown in Figure 6. The diagram is shown in Figure 7.

Steel balls (with diameter of 34 mm and mass of 160.5 g) were used for the test. A steel ball was chosen due to the significant difference in stiffness between the tested material and steel. In order to reduce reflections on the ball, which would introduce contamination to the image (and the position of the ball), the ball was painted with matte black paint.

The height from which the ball was freely dropped was 180 mm. This height was chosen due to the capabilities of the camera lens and the ability to measurably observe the ball throughout its entire course. Another argument for this drop height was the possibility of recording the ball's indentation in the material (local minimum of the ball's position). In the case of preliminary tests at lower drop heights, there were problems with registering the local minimum. At higher altitudes, the speed of the ball falling right after the collision was so high that the error in determining the position of the ball at a given number of frames of the recorded image increased significantly.

Figure 6. Test bench for ball-tracking experiment.

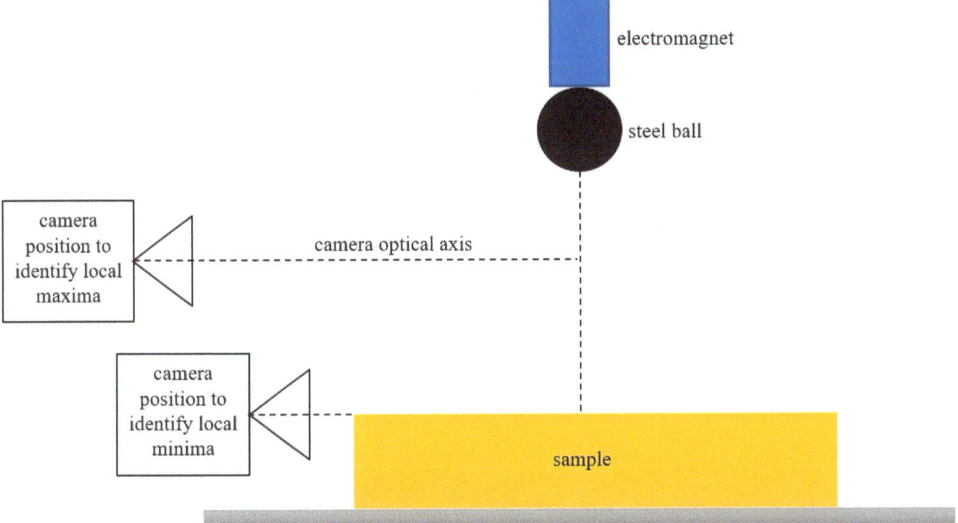

Figure 7. Test bench diagram.

The choice of two camera positions—close and far (15 cm and 40 cm away from the center of the ball, respectively)—was dictated by two requirements. The first one was to enable the recording of the first reflections and the beginning of the drop. The second was a more accurate recording of the indentation. Although the far position allowed us to achieve these two requirements with one camera position, a close position was also used to increase the accuracy of the indentation analysis.

Six ball drops were performed for one sample. This decision was made to increase the robustness of the results by determining the average ball path. After each drop, it was verified if any local yielding occurred at the point of contact between the sample and the ball. There were no permanent deformations of the tested materials.

As the last step, indentation was calculated based on ball tracking. Indentation was defined as the maximum absolute value of the ball bottom position during contact with sample (local minimum values are presented in Figure 8). Indentation was determined for each of the observed contacts. The rebound height was determined as local maximum values of ball tracking (with ball bottom position).

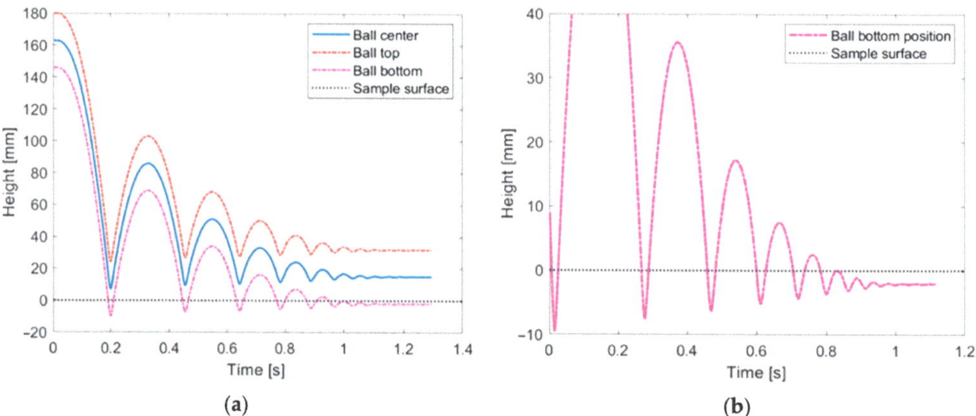

Figure 8. Examples of ball-tracking results for the P150 sample with a density of 288 kg/m^3. (**a**) shows full motion of the ball, and (**b**) shows magnification in ball drop area.

3. Results

The results for the reference method are presented in Table 3. The presented results are a summary of the averaged values of the critical damping factor (CDF), resonant frequency (Rf), and dynamic stiffness (DS) for each of the tested materials. Additionally, the values for the 95% confidence intervals (95%CI) are presented in parentheses.

Table 3. Results showing the critical damping factor, resonant frequency, and dynamic stiffness of the tested samples obtained from the reference method.

Material Name	CDF (−) (95%CI)	Rf (Hz) (95%CI)	DS (MN/m^3) (95%CI)
rubber granulate with rubber fibers (F570)	0.0675 (0.0581, 0.0769)	73.28 (69.24, 77.31)	42.4 (37.9, 47)
rubber granulate with rubber fibers (F700)	0.0674 (0.064, 0.0709)	77.61 (74.32, 80.9)	47.6 (43.6, 51.6)
rebound polyurethane (P150)	0.0691 (0.0501, 0.0881)	35.5 (33.8, 37.2)	10.0 (9.0, 10.9)
rebound polyurethane (P250)	0.0805 (0.0699, 0.0911)	69.97 (64.93, 75.02)	38.7 (33.1, 44.3)
rubber granulate (S1000)	0.1127 (0.1023, 0.1231)	76.31 (70.18, 82.43)	46.1 (38.7, 53.4)
rubber granulate (S850)	0.1123 (0.0962, 0.1284)	84.39 (78.88, 89.89)	56.3 (49, 63.6)

The post-processed and averaged results of tracing the ball during free fall are presented in Tables 4 and 5. Table 4 shows the results of the relative bounce height up to the

4th bounce. The relative bounce height was calculated in the following manner: The first relative rebound height was the ratio of the start height to the first rebound height (1st peak/start). The second relative rebound height was the ratio of the second to the first rebound height (2nd/1st peak). The third relative rebound height was the ratio of the third to the second rebound height (3rd/2nd peak). And the last and fourth rebound height was calculated in the same manner. Additionally, the values for the 95% confidence intervals (95%CI) of the calculated relative rebound heights are presented in parentheses.

Table 4. Results showing relative bounce height values obtained from ball tracking.

	1st Peak/Start (95%CI)	2nd/1st Peak (95%CI)	3rd/2nd Peak (95%CI)	4th/3rd Peak (95%CI)
rubber granulate with rubber fibers (F570)	0.4877 (0.4703, 0.5052)	0.5134 (0.4995, 0.5272)	0.5091 (0.4985, 0.5197)	0.497 (0.4845, 0.5095)
rubber granulate with rubber fibers (F700)	0.5183 (0.5027, 0.5338)	0.5549 (0.5356, 0.5743)	0.5595 (0.5485, 0.5705)	0.5547 (0.5455, 0.564)
rebound polyurethane (P150)	0.4764 (0.4625, 0.4903)	0.4973 (0.4893, 0.5053)	0.4819 (0.4623, 0.5014)	0.4455 (0.4271, 0.4639)
rebound polyurethane (P250)	0.4935 (0.4871, 0.4999)	0.5199 (0.5103, 0.5295)	0.5158 (0.5063, 0.5253)	0.4981 (0.4845, 0.5116)
rubber granulate (S1000)	0.3701 (0.3353, 0.405)	0.3763 (0.3366, 0.4159)	0.3555 (0.3299, 0.3812)	0.363 (0.3131, 0.4128)
rubber granulate (S850)	0.3957 (0.3259, 0.4654)	0.4183 (0.3648, 0.4718)	0.4114 (0.3703, 0.4525)	0.3902 (0.329, 0.4513)

Table 5. Results showing relative indentation values obtained from tracking the ball's path.

	1st Ind. */Start (95%CI)	2nd/1st Ind. (95%CI)	3rd/2nd Ind. (95%CI)	4th/3rd Ind. (95%CI)
rubber granulate with rubber fibers (F570)	0.0201 (0.0187, 0.0215)	0.0281 (0.0257, 0.0304)	0.0373 (0.0335, 0.0412)	0.0451 (0.0406, 0.0496)
rubber granulate with rubber fibers (F700)	0.0177 (0.0167, 0.0187)	0.0263 (0.0223, 0.0303)	0.0323 (0.0241, 0.0404)	0.0382 (0.027, 0.0494)
rebound polyurethane (P150)	0.0547 (0.0505, 0.0589)	0.0825 (0.0778, 0.0872)	0.1132 (0.1002, 0.1263)	0.1375 (0.1218, 0.1533)
rebound polyurethane (P250)	0.0293 (0.0252, 0.0335)	0.0415 (0.036, 0.0469)	0.0547 (0.0453, 0.064)	0.0675 (0.0567, 0.0782)
rubber granulate (S1000)	0.0084 (0.0064, 0.0104)	0.0172 (0.0136, 0.0208)	0.0269 (0.0205, 0.0333)	0.0348 (0.0229, 0.0467)
rubber granulate (S850)	0.013 (0.0108, 0.0152)	0.0245 (0.0161, 0.0329)	0.0378 (0.0249, 0.0507)	0.0479 (0.0324, 0.0634)

* Ind. means indentation.

Table 5 shows the results for the averaged and post-processed relative indentations. Similar to the results with the relative rebound height, it was decided to show the results only up to the 4th indentation. The relative indentation was calculated in the following manner: The first relative indentation (1st ind.*/start) was the ratio of the first indentation depth to the start position of the ball bottom. The second relative indentation (2nd/1st ind.) was the ratio of the second indentation depth to the first rebound height with the ball bottom position. The third and forth relative indentations were calculated in the same way. Additionally, the values for the 95% confidence intervals (95%CI) of the calculated relative indentations are presented in parentheses.

In the next part of the paper, an examination of any relationships between the reference method for testing dynamic stiffness and the relative values measured by ball tracking was performed.

4. Discussion

First of all, the question must be raised as to if the alternative method can be related to the reference method in any way. The reference method has its basis in the literature [100,115] and ISO standard [48].

4.1. Dependencies between Relative Indentation and Dynamic Stiffness

As a first topic, it can be discussed if the indentation can provide direct information about the stiffness of the material. The indentation itself strongly depends on the height of the drop; hence, the decision was made to use relative indentation, which shows the ratio of the rebound height of subsequent indices. While examining various mathematical models of the relationship between the relative indentation and dynamic stiffness, it was decided to choose a linear model using a linear function. Goodness of fit is presented in Table 6.

Table 6. Initial goodness of fit results for val(x) = p1·x + p2 function, where x is relative indentation and val(x) is dynamic stiffness.

	1st Ind./Start	2nd/1st Ind.	3rd/2nd Ind.	4th/3rd Ind.	5th/4th Ind.
SSE	747.6245	788.0932	911.2817	969.5523	1059.64
R^2	0.858681	0.851032	0.827746	0.816732	0.799703
RMSE	5.829488	5.985183	6.435984	6.638566	6.940134

From the data included in Table 6 it can be concluded that the first relative indentation to dynamic stiffness correlates best. However, this raises some doubts, because a linear function has a zero crossing (unless it is constantly positive) and also accepts negative arguments (which would mean that relative indentation can be negative). For this reason, it is worth considering another function describing the relationship between relative indentation and dynamic stiffness. As an alternative to the linear function, one can propose the rational function or power function, which is characterized by asymptotic behavior. The analyses show a better R^2 for the power function, which was (however) still not satisfactory (R^2~0.5). The matching results are shown in Figure 9.

Although the power function has a physical justification—it does not allow negative dynamic stiffness or relative indentation—the R^2 for the linear function is much higher (R^2~0.54 to R^2~0.86, respectively). It can therefore be said that in a limited range of relative indentation (0.01 to 0.06), fitting with a linear function gives sufficient accuracy.

Focusing on the results, it can be noticed that in the relative indentation near value 0.01, there is a large dispersion of the results for dynamic stiffness. These are the results for samples S1000 (lowest relative indentation) and S850 (relative indentation ~0.015). There may be several reasons for this condition. The first reason is the apparent softening of the S1000 samples due to a lack of full contact with the pressure plate due to their granulation. The second reason may be limitations related to the accuracy of recording small indentations. Although the mentioned reasons constituting limitations of the methods used are not the subject of this paper, it is worth taking a closer look at this problem in order to draw conclusions for further studies.

Therefore, it was decided to remove the S1000 samples from the analysis and repeat the previously presented analysis. The results are shown in Figure 10.

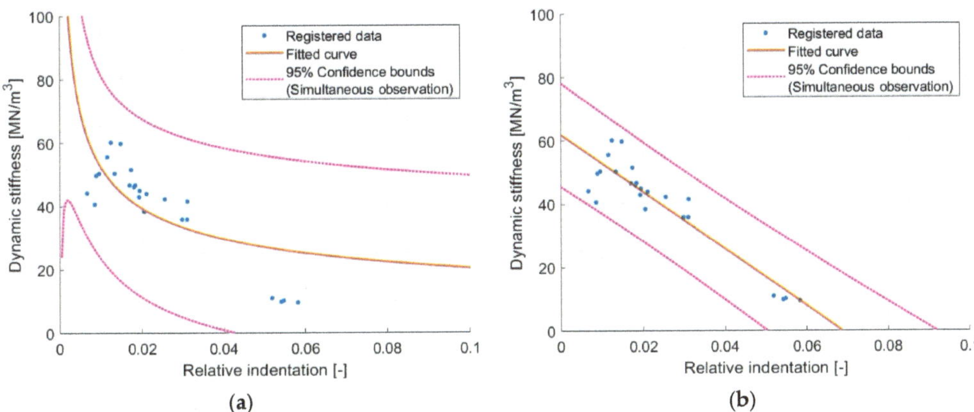

Figure 9. Comparison of different models of dynamic stiffness and relative indentation relationship. (**a**) model Power1 val(x) = a·x^b, where a = 7.834, b = −0.4115, and R^2 = 0.5435 and (**b**) model Linear val(x) = p1·x + p2, where p1 = −899.6, p2 = 61.65, and R^2 = 0.8587.

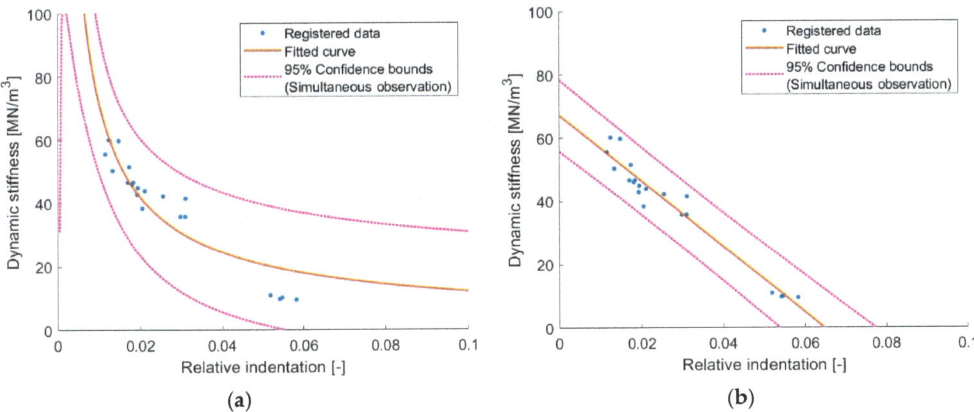

Figure 10. Comparison of different models of dynamic stiffness and relative indentation relationship with exclusion of S1000 (relative indentation < 0.01). (**a**) model Power1 val(x) = a·x^b, where a = 2.000, b = −0.777, and R^2 = 0.8381 and (**b**) model Linear val(x) = p1·x + p2, where p1 = −1036, p2 = 66.95, and R^2 = 0.9468.

After removing the S1000 samples from the analysis, a significant improvement in correlation was obtained for both models. For the power function, an increase in R^2 was obtained from ~0.55 to ~0.85, which allowed us to demonstrate an acceptable correlation. For the linear function, an increase in R^2 was obtained from ~0.85 to ~0.95, which indicates a very good correlation.

In conclusion, it can be indicated that it is possible to determine dynamic stiffness by examining relative indentation. However, the obtained results indicate that this possibility exists to a limited extent in the relative indentation (0.01–0.06) for the tested sample geometry. Attempts to use possible extrapolation using the power and linear functions do not provide appropriate results.

4.2. Dependencies between Relative Rebound and Critical Damping Factor

Damping is the second parameter examined in this paper. In the case of damping, it is expected that the relative height to which the ball bounces will be correlated with the critical damping factor. This is due to the dissipation of the total energy of the ball by the sample during the rebound process.

As in the case of relative indentation, various models were tested for the relationship between the critical damping factor and the relative rebound. Yet again, it was noticed that the linear function gives the highest R^2 for the recorded measurements. The results of this correlation for the subsequent relative rebound are presented in Table 7.

Table 7. Initial goodness of fit results for val(x) = p1·x + p2 function, where x is relative rebound and y is critical damping factor.

	1st Peak/Start	2nd/1st Peak	3rd/2nd Peak	4th/3rd Peak
SSE	0.001982	0.002138	0.002646	0.003137
R^2	0.814922	0.800411	0.752947	0.707063
RMSE	0.009492	0.009857	0.010967	0.011942

In the case of the first and second relative rebounds, a high R^2 (~0.8) was observed, while in further relative rebounds, a significant decrease in the R^2 value was observed.

Relative rebound, similar to relative indentation, correlates well with the linear function. However, this does not change the fact that there is no physical justification for such a model. Hence, there is a need to propose another, more rational function. It is known that in the theoretical case, when the material has no damping, that the ball will bounce endlessly. Naturally, this was assuming no air resistance and no damping in the ball itself. Hence, at a relative rebound equal to 1, the CDF should be equal to 0. Moreover, when increasing the CDF for the sample, the relative rebound decreases. In theory, the material may have damping above the critical damping, but in practice, such materials are not used.

Therefore, two functions that can describe the dependencies between the CDF and the relative rebound have been proposed—linear and logarithmic functions. The match results for the first relative rebound are shown in Figure 11.

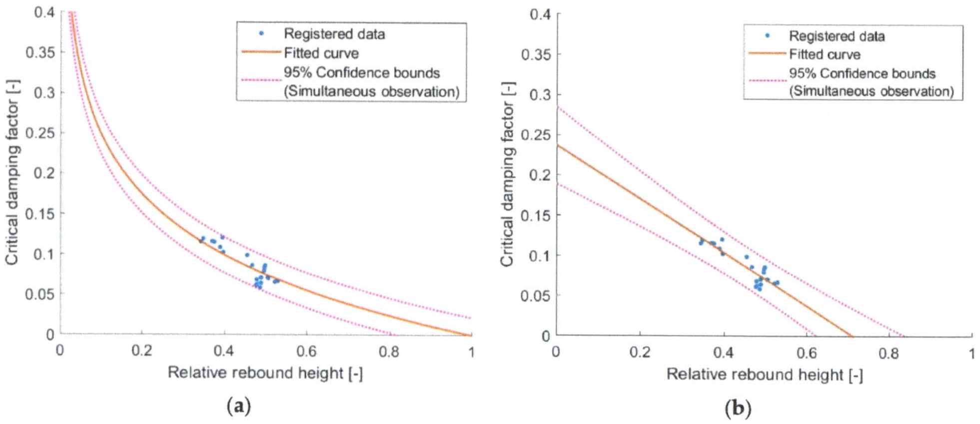

Figure 11. Comparison of different models of critical damping factor and first relative rebound relationship. (a) model Logarithmic val(x) = a·log10(x), where a = −0.2492 and R^2 = 0.7654 and (b) model Linear val(x) = p1·x + p2, where p1 = −0.3325, p2 = 0.2368, and R^2 = 0.8149.

In the case of a logarithmic function that preserves the boundary condition associated with the lack of damping, R^2~0.77 was demonstrated, which means an acceptable correlation. For the best-fitting linear function, R^2 is ~0.81, which already means a good correlation. However, the R^2 values are comparable with a slight advantage of the linear function.

Taking into account the fact that for the results presented in the Table 7, the R^2 was ~0.8 for the 1st and 2nd relative rebounds, it was decided to average the results for these two relative rebounds and make the same adjustment again. The results are shown in Figure 12.

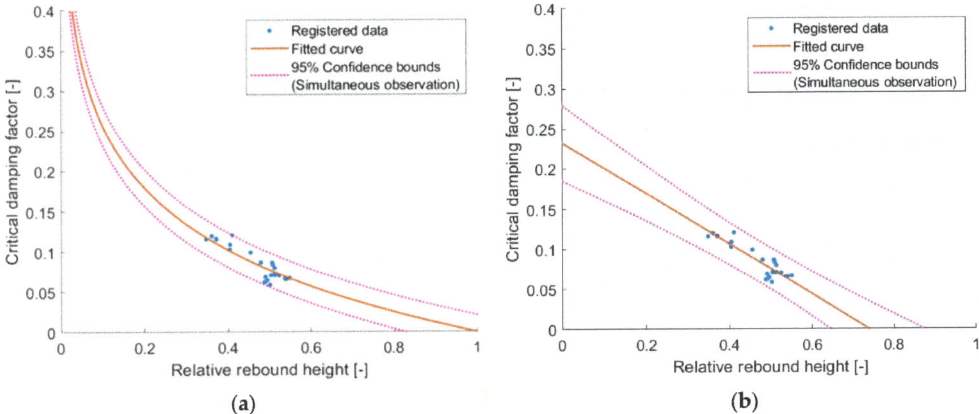

Figure 12. Comparison of different models of critical damping factor and first and second relative rebound (averaged) relationship. (**a**) model Logarithmic val(x) = a·log10(x), where a = −0.2568 and R^2 = 0.7809 and (**b**) model Linear val(x) = p1·x + p2, where p1 = −0.3143, p2 = 0.2321, and R^2 = 0.8123.

For the logarithmic function, an increase in R^2 was observed from 0.765 to 0.781. For the linear function, there was a slight decrease in R^2 from 0.815 to 0.812. Therefore, it can be said that both models correctly predict the behavior of the dependence of the CDF on the relative rebound. It is worth mentioning that the model using the logarithmic function allows for some extrapolation of the results.

4.3. Reference Method—Density, Dynamic Stiffness, and Critical Damping Factor

Using the results from the reference method for analysis, we checked if there were any relationships between the measured values for the tested samples. Considering that the tests were carried out using materials with different internal structures, and for which the only common feature was their use as passive vibration isolation, having such trivial data as density, it would be beneficial to estimate their other mechanical parameters. However, strong correlations should not be expected, because material density is one of many parameters responsible for stiffness or damping.

Various models were tested to examine the dependence of dynamic stiffness on density and the critical damping factor on density. The highest R^2 in both cases was obtained for the power function summed with the free term. Figure 13 shows the results of the discussed adjustments.

For the tested set of samples, describing the relationship between dynamic stiffness and density gives an R^2 of 0.884. Although this would indicate a good correlation, it is difficult to speak of a more general principle. Taking into account the 95% confidence bounds, it turns out that according to the proposed model, for a density of, e.g., 800 kg/m^3, the dynamic stiffness results are between ~30 MN/m^3 and ~65 MN/m^3. Consequently, even for engineering applications, this is a rather rough estimate of dynamic stiffness. The reason for this arrangement can be seen in the fact that rubbers dominate in high densities and rebound polyurethanes in low densities. On the other hand, we have the problematic S1000 material, which appears to have lower dynamic stiffness compared to less dense materials of a similar type.

With the dependence of the CDF on density, an R^2 of 0.630 was obtained. Here, R^2 shows a reduced ability to make the CDF predictions based on density. Of course, the problems indicated above could be discussed, but the graph shows that the attenuation increases significantly only at higher densities. When the large scatter at the 95% confidence bounds is ignored, it can be concluded that the attenuation increases with the density of the material.

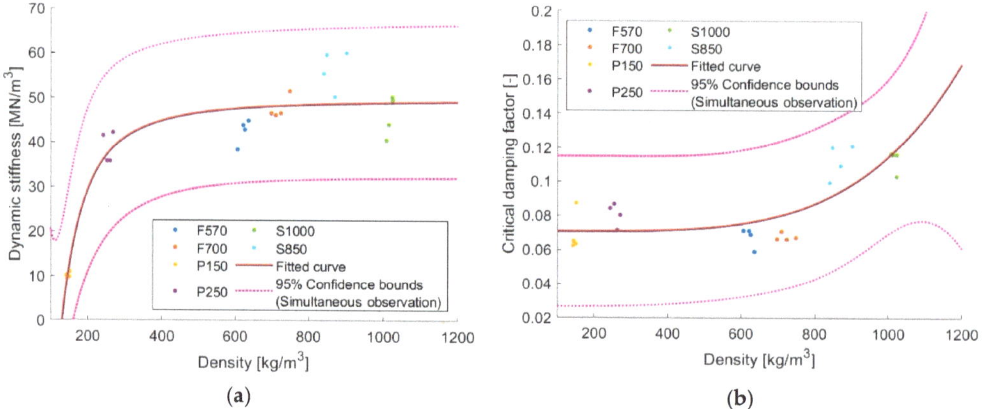

Figure 13. Models of Power2 (val(x) = a·x^b + c) describing relationship between (**a**) dynamic stiffness and density where a = -1.422×10^6, b = -2.105, c = 49.53, and R^2 = 0.8842 and (**b**) critical damping factor and density where a = 2.201×10^{-15}, b = 4.433, c = 0.07053, and R^2 = 0.6301.

Bearing in mind that in the tested materials the increase in dynamic stiffness with density occurs rather at the beginning of the analyzed density and the damping increases in the second half of this range, an attempt can be made to find a resultant relationship. It was checked if the CDF/DS ratio depends to some extent on density. The proposed model was a linear combination of two exponential functions. The result is presented in Figure 14.

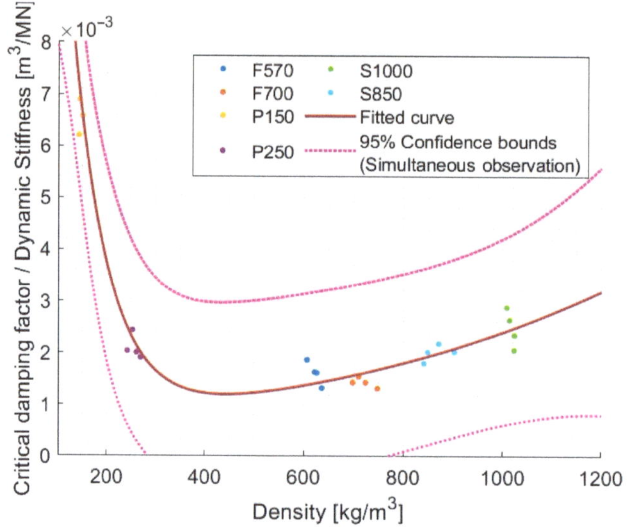

Figure 14. The fraction Critical damping factor/Dynamic stiffness as a function of density described with the usage of Exp2 model (val(x) = a·exp(b·x) + c·exp(d·x)), a = 0.04442, b = -0.01349, c = 0.0005713, d = 0.001433, and R^2 = 0.9460.

The value of R^2 = 0.946 indicates a very good correlation. For rubber materials, regardless of their internal structure—whether it is just granules or granules with fibers—a constant increase is observed. For rebound polyurethane, the analyzed value quickly decreases depending on the density. This allowed us to conclude that although there is no direct relationship between the dynamic stiffness or the CDF and density, there is an indication that the ratio of CDF/DS may depend directly on the density even for different materials. It is definitely worth further study.

4.4. Reference Method—Rayleigh Damping

Rayleigh damping is widely used in computational methods [119–121] due to its simplicity and computational efficiency. Based on data from various materials, it was decided to check if Rayleigh damping can be considered on a dynamic stiffness testing machine. The results are shown in Figure 15.

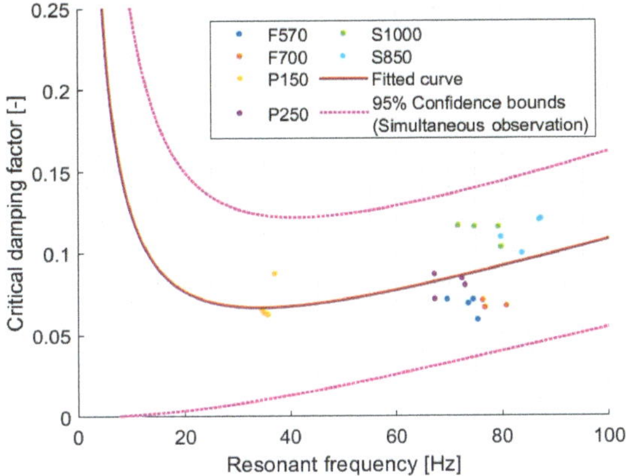

Figure 15. Rayleigh damping model (val(x) = 1/2·(a·x + b/x)), where a = 0.001935, b = 2.273, and R^2 = 0.2385.

In the case of the tested machine, for various materials subjected to the test procedure, no significant relationship was found that would indicate that there is Rayleigh damping in the form described in its definition, because R^2 = 0.239 indicates the lack of a relationship.

As an alternative to pure Rayleigh damping, it was checked if Rayleigh's relationship between the CDF and the DS could be used. The results of this adjustment are shown in Figure 16.

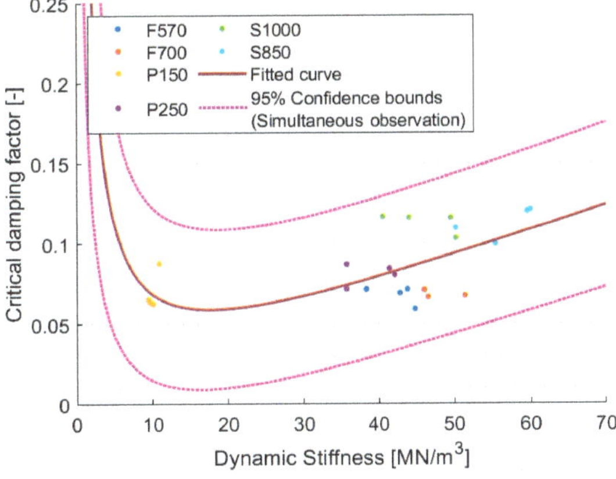

Figure 16. Rayleigh-ish damping model based on dynamic stiffness instead of resonant frequency (val(x) = 1/2·(a·x + b/x)), where a = 0.003327, b = 1.029, and R^2 = 0.3730.

Although a slightly higher $R^2 = 0.373$ was obtained, it indicated a lack of a relationship in this case as well.

5. Conclusions

General Conclusions and Further Studies

The following conclusions can be drawn from the research conducted for this paper.

The method using relative rebound and relative indentation gives a good prediction of the dynamic stiffness and critical damping factor parameters. Taking into account that the R^2 in all models discussed is above 0.7, and in the case of certain limitations even above 0.9, it confirms that the alternative method of using image analysis is an effective estimate of material parameters.

It turns out that only the initial measurement (1st or eventually 2nd rebound) can be used for research. Moreover, when examining the relationship, it was in this region that the highest R^2 values were obtained, which allows for the additional shortening of the test time without losing its accuracy. Very weak dependencies of dynamic stiffness on density and the critical damping factor on density were found. This makes it impossible to predict the above-mentioned parameters based on density. However, a certain trend can be identified where dynamic stiffness and damping increase with density.

An interesting observation is the strong relationship between the critical damping factor/dynamic stiffness fraction and material density. The R^2 value is very high (~0.95). It can be concluded that for low densities, the damping is very effective per unit of dynamic stiffness. This efficiency drops to its minimum at approximately $400\ \text{kg/m}^3$ and then gradually increases.

There is no strong relationship in the case of tests on a machine compliant with ISO 9052 [48] that would allow us to identify Rayleigh damping. In the case of the model using dynamic stiffness instead of resonant frequency, a certain trend can be noticed, showing an increase in the critical damping factor with the increase in dynamic stiffness.

The main disadvantage of an assessment using image processing is its accuracy. Although R^2 values around 0.8 for the critical damping factor prediction and R^2 values around 0.9 were obtained for the tested samples, the extrapolation of these results beyond the framework specified in the article for dynamic stiffness (at 10–$60\ \text{MN/m}^3$) and damping (for 6–12%) is problematic. Particular attention should be focused on the S1000 sample, which, although slightly lower in dynamic stiffness, showed lower relative indentation than the proposed models would indicate. It is worth emphasizing that the surface structure of the sample has a significant impact on the relative indentation. Samples with a large structure (large pore size on the surface) may appear softer when tested because the ball sinks deeper during contact.

The observed discrepancies related to the decrease in stiffness with increasing density for the S1000 material prompted an in-depth examination of the causes of this phenomenon. It is suspected that this may be related to the reduction in the contact surface of the pressure plate of the dynamic stiffness testing machine. The reason may also be found in the technology of making granules and the possibility to use a more resilient glue or reducing its amount compared to other granulates. Another way of distributing the granules can also be indicated. At this time, the cause of this condition is not clear; therefore, additional tests should be performed, taking into account the above considerations.

It is worth mentioning that ball tracking was performed using an algorithm that directly tracks the ball's path. However, it is worth considering much more complex methods. One such method is Eulerian video magnification [122,123], which could allow for more accurate measurements, especially of the contact phenomenon during the free fall of the ball.

Author Contributions: Conceptualization, K.N. (Krzysztof Nering) and K.N. (Konrad Nering); methodology, K.N. (Krzysztof Nering) and K.N. (Konrad Nering); software, K.N. (Krzysztof Nering) and K.N. (Konrad Nering); validation, K.N. (Krzysztof Nering) and K.N. (Konrad Nering); formal analysis, K.N. (Krzysztof Nering) and K.N. (Konrad Nering); investigation, K.N. (Krzysztof Nering) and K.N. (Konrad Nering); resources, K.N. (Krzysztof Nering) and K.N. (Konrad Nering); data curation, K.N. (Krzysztof Nering) and K.N. (Konrad Nering); writing—original draft preparation, K.N. (Krzysztof Nering) and K.N. (Konrad Nering); writing—review and editing, K.N. (Krzysztof Nering) and K.N. (Konrad Nering); visualization, K.N. (Krzysztof Nering) and K.N. (Konrad Nering); supervision, K.N. (Krzysztof Nering) and K.N. (Konrad Nering); project administration, K.N. (Krzysztof Nering) and K.N. (Konrad Nering); funding acquisition, K.N. (Krzysztof Nering) and K.N. (Krzysztof Nering). All authors have read and agreed to the published version of the manuscript.

Funding: This research received no external funding.

Institutional Review Board Statement: Not applicable.

Informed Consent Statement: Not applicable.

Data Availability Statement: Data are contained within the article.

Acknowledgments: The article was written based on the authors' knowledge gained through work on the Vibroacoustic Testing Line PM&VL 7.3 "Vibro-acoustic comfort subjected to noise and vibrations—connectors influence" in the MEZeroE Measuring Envelope project for products and systems contributing to healthy, nearly zero-energy buildings, which received funding from EU Horizon 2020 research and innovation program under subsidy No. 953157.

Conflicts of Interest: The authors declare no conflicts of interest.

References

1. Liu, S.; Jia, L.; Zhang, F.; Wang, R.; Liu, X.; Zou, L.; Tang, X. Do New Urbanization Policies Promote Sustainable Urbanization? Evidence from China's Urban Agglomerations. *Land* **2024**, *13*, 412. [CrossRef]
2. Abulimiti, A.; Liu, Y.; Yang, L.; Abulikemu, A.; Mamitimin, Y.; Yuan, S.; Enwer, R.; Li, Z.; Abuduaini, A.; Kadier, Z. Urbanization Effect on Changes in Extreme Climate Events in Urumqi, China, from 1976 to 2018. *Land* **2024**, *13*, 285. [CrossRef]
3. Boanada-Fuchs, A.; Kuffer, M.; Samper, J. A Global Estimate of the Size and Location of Informal Settlements. *Urban. Sci.* **2024**, *8*, 18. [CrossRef]
4. Ge, S.; Zhan, W.; Wang, S.; Du, H.; Liu, Z.; Wang, C.; Wang, C.; Jiang, S.; Dong, P. Spatiotemporal Heterogeneity in Global Urban Surface Warming. *Remote Sens. Environ.* **2024**, *305*, 114081. [CrossRef]
5. Bian, H.; Gao, J.; Liu, Y.; Yang, D.; Wu, J. China's Safe and Just Space during 40 Years of Rapid Urbanization and Changing Policies. *Landsc. Ecol.* **2024**, *39*, 74. [CrossRef]
6. Turemuratov, O.; Byulegenova, B.; Pogodin, S.; Onuchko, M.; Nurtazina, R. Urbanization Trends in Central Asian Countries: Aspects of Extensive and Intensive Agglomeration Growth. *Public Organ. Rev.* **2024**. [CrossRef]
7. Chanieabate, M.; He, H.; Guo, C.; Abrahamgeremew, B.; Huang, Y. Examining the Relationship between Transportation Infrastructure, Urbanization Level and Rural-Urban Income Gap in China. *Sustainability* **2023**, *15*, 8410. [CrossRef]
8. Mwanzu, A.; Nguyu, W.; Nato, J.; Mwangi, J. Promoting Sustainable Environments through Urban Green Spaces: Insights from Kenya. *Sustainability* **2023**, *15*, 11873. [CrossRef]
9. Aguilera, M.A.; González, M.G. Urban Infrastructure Expansion and Artificial Light Pollution Degrade Coastal Ecosystems, Increasing Natural-to-Urban Structural Connectivity. *Landsc. Urban. Plan.* **2023**, *229*, 104609. [CrossRef]
10. Dalla Longa, R. Urban Infrastructures. In *Urban Infrastructure*; Springer International Publishing: Cham, Switzerland, 2023; pp. 1–42.
11. Nicoletti, L.; Sirenko, M.; Verma, T. Disadvantaged Communities Have Lower Access to Urban Infrastructure. *Environ. Plan. B Urban. Anal. City Sci.* **2023**, *50*, 831–849. [CrossRef]
12. Logachev, M.; Zhukova, G. Parametric Modeling of Traffic Flows for Predicting Sustainable Development of Urban Areas. *BIO Web Conf.* **2024**, *83*, 05002. [CrossRef]
13. Martínez-Corral, A.; Cárcel-Carrasco, J.; Fonseca, F.C.; Kannampallil, F.F. Urban–Spatial Analysis of European Historical Railway Stations: Qualitative Assessment of Significant Cases. *Buildings* **2023**, *13*, 226. [CrossRef]
14. Ahac, M.; Ahac, S.; Majstorović, I.; Stepan, Ž. Contribution to Rail System Revitalization, Development, and Integration Projects Evaluation: A Case Study of the Zadar Urban Area. *Infrastructures* **2024**, *9*, 32. [CrossRef]
15. Megna, G.; Governi, L.; Carfagni, M.; Borchi, F.; Puggelli, L.; Buonamici, F. State of the Art about Solutions for Tram Noise Reduction in the Framework of the Life SNEAK Project. In Proceedings of the INTER-NOISE and NOISE-CON Congress and Conference Proceedings, Grand Rapids, MI, USA, 15–18 May 2023; Volume 265, pp. 6302–6313. [CrossRef]
16. Khan, I.; Khattak, K.; Khan, Z.H.; Gulliver, T.A. Impact of Road Pavement Condition on Vehicular Free Flow Speed, Vibration and In-Vehicle Noise. *Sci. Eng. Technol.* **2023**, *3*, 1–8. [CrossRef]
17. Khan, D.; Burdzik, R. Measurement and Analysis of Transport Noise and Vibration: A Review of Techniques, Case Studies, and Future Directions. *Measurement* **2023**, *220*, 113354. [CrossRef]
18. Liu, Q.; Xu, L.; Feng, Q. Intelligent Monitoring of Vibration and Structural-Borne Noise Induced by Rail Transit. *Intell. Transp. Infrastruct.* **2023**, *2*, liad013. [CrossRef]

19. Kawecki, J.; Kowalska, A. Analysis of Vibration Influence on People in Buildings in Standards Approach. *WIT Trans. Ecol. Environ.* **2011**, *148*, 355–366.
20. Kowalska-Koczwara, A. Influence of Location of Measurement Point on Evaluation of Human Perception of Vibration. *J. Meas. Eng.* **2019**, *7*, 147–154. [CrossRef]
21. Nering, K.; Kowalska-Koczwara, A.; Stypuła, K. Annoyance Based Vibro-Acoustic Comfort Evaluation of as Summation of Stimuli Annoyance in the Context of Human Exposure to Noise and Vibration in Buildings. *Sustainability* **2020**, *12*, 9876. [CrossRef]
22. Fedorczak-Cisak, M.; Kowalska-Koczwara, A.; Nering, K.; Pachla, F.; Radziszewska-Zielina, E.; Śladowski, G.; Tatara, T.; Ziarko, B. Evaluation of the Criteria for Selecting Proposed Variants of Utility Functions in the Adaptation of Historic Regional Architecture. *Sustainability* **2019**, *11*, 1094. [CrossRef]
23. Güneralp, B.; Reba, M.; Hales, B.U.; Wentz, E.A.; Seto, K.C. Trends in Urban Land Expansion, Density, and Land Transitions from 1970 to 2010: A Global Synthesis. *Environ. Res. Lett.* **2020**, *15*, 044015. [CrossRef]
24. Chakraborty, S.; Maity, I.; Dadashpoor, H.; Novotn, J.; Banerji, S. Building in or out? Examining Urban Expansion Patterns and Land Use Efficiency across the Global Sample of 466 Cities with Million+ Inhabitants. *Habitat. Int.* **2022**, *120*, 102503. [CrossRef]
25. Amen, M.A.; Afara, A.; Nia, H.A. Exploring the Link between Street Layout Centrality and Walkability for Sustainable Tourism in Historical Urban Areas. *Urban. Sci.* **2023**, *7*, 67. [CrossRef]
26. McDonald, R.I.; Aronson, M.F.J.; Beatley, T.; Beller, E.; Bazo, M.; Grossinger, R.; Jessup, K.; Mansur, A.V.; Puppim de Oliveira, J.A.; Panlasigui, S.; et al. Denser and Greener Cities: Green Interventions to Achieve Both Urban Density and Nature. *People Nat.* **2023**, *5*, 84–102. [CrossRef]
27. He, X.; Yu, Y.; Jiang, S. City Centrality, Population Density and Energy Efficiency. *Energy Econ.* **2023**, *117*, 106436. [CrossRef]
28. Goda, T.; Stewart, C.; Torres García, A. Absolute Income Inequality and Rising House Prices. *Socioecon. Rev.* **2021**, *18*, 941–976. [CrossRef]
29. Kettunen, H.; Ruonavaara, H. Rent Regulation in 21st Century Europe. Comparative Perspectives. *Hous. Stud.* **2021**, *36*, 1446–1468. [CrossRef]
30. Manting, D.; Kleinepier, T.; Lennartz, C. Housing Trajectories of EU Migrants: Between Quick Emigration and Shared Housing as Temporary and Long-Term Solutions. *Hous. Stud.* **2024**, *39*, 1027–1048. [CrossRef]
31. Andargie, M.S.; Touchie, M.; O'Brien, W. Case Study: A Survey of Perceived Noise in Canadian Multi-Unit Residential Buildings to Study Long-Term Implications for Widespread Teleworking. *Build. Acoust.* **2021**, *28*, 443–460. [CrossRef]
32. Lee, P.J.; Jeong, J.H. Attitudes towards Outdoor and Neighbour Noise during the COVID-19 Lockdown: A Case Study in London. *Sustain. Cities Soc.* **2021**, *67*, 102768. [CrossRef]
33. Deaconu, M.; Cican, G.; Cristea, L. Noise Impact Mitigation of Shopping Centres Located near Densely Populated Areas for a Better Quality of Life. *Appl. Sci.* **2020**, *10*, 6484. [CrossRef]
34. Directive 2002/49/EC of the European Parliament and of the Council of 25 June 2002 Relating to the Assessment and Management of Environmental Noise—Declaration by the Commission in the Conciliation Committee on the Directive Relating to the Assessment and Management of Environmental Noise. Available online: https://eur-lex.europa.eu/LexUriServ/LexUriServ.do?uri=CELEX:32002L0049:EN:HTML (accessed on 1 December 2022).
35. Noise Control Act of 1972, P.L. 92-574, 86 Stat. 1234. Available online: https://www.gsa.gov/system/files/Noise_Control_Act_of_1972.pdf (accessed on 1 December 2022).
36. *ISO 9612:2009*; Acoustics—Determination of Occupational Noise Exposure—Engineering Method. ISO: Geneva, Switzerland, 2009.
37. *ISO 2631-1:1997*; Evaluation of Human Exposure to Whole-Body Vibration. ISO: Geneva, Switzerland, 1997.
38. *ISO 16032:2004*; Acoustics—Measurement of Sound Pressure Level from Service Equipment in Buildings—Engineering Method. ISO: Geneva, Switzerland, 2004.
39. *ISO 2631-1:1985*; Evaluation of Human Exposure to Whole-Body Vibration Part 1: General Requirements. ISO: Geneva, Switzerland, 1958.
40. *ISO 12354-2:2017*; Building Acoustics—Estimation of Acoustic Performance of Buildings from the Performance of Elements—Part 2: Impact Sound Insulation between Rooms. ISO: Geneva, Switzerland, 2017.
41. *ISO 12354-1:2017*; Building Acoustics—Estimation of Acoustic Performance of Buildings from the Performance of Elements—Part 1: Airborne Sound Insulation between Rooms. ISO: Geneva, Switzerland, 2017.
42. *ISO 6897:1984*; Guidelines for the Evaluation of the Response of Occupants of Fixed Structures, Especially Buildings and off-Shore Structures, to Low Frequency Horizontal Motion (0.063 to 1 Hz). ISO: Geneva, Switzerland, 1984.
43. Howarth, H.V.C.; Griffin, M.J. The Annoyance Caused by Simultaneous Noise and Vibration from Railways. *J. Acoust. Soc. Am.* **1991**, *89*, 2317–2323. [CrossRef]
44. Howarth, H.V.C.; Griffin, M.J. The Relative Importance of Noise and Vibration from Railways. *Appl. Ergon.* **1990**, *21*, 129–134. [CrossRef]
45. Öhrström, E.; Skånberg, A.B. A Field Survey on Effects of Exposure to Noise and Vibration from Railway Traffic, Part I: Annoyance and Activity Disturbance Effects. *J. Sound. Vib.* **1996**, *193*, 39–47. [CrossRef]
46. Wrótny, M.; Bohatkiewicz, J. Impact of Railway Noise on People Based on Strategic Acoustic Maps. *Sustainability* **2020**, *12*, 5637. [CrossRef]

47. Kowalska-Koczwara, A.; Pachla, F.; Nering, K. Environmental Protection Against Noise and Vibration. *IOP Conf. Ser. Mater. Sci. Eng.* **2021**, *1203*, 032026. [CrossRef]
48. *EN 29052-1:2011/ISO 9052-1*; Acoustics—Determination of Dynamic Stiffness—Part 1: Materials Used under Floating Floors in Dwellings. Available online: https://Standards.Iteh.Ai/Catalog/Standards/Cen/04a2dad0-Ff01-4409-B2d4-6d6d62feb5c7/En-29052-1-1992 (accessed on 1 December 2022).
49. Kumar, D.; Alam, M.; Zou, P.X.W.; Sanjayan, J.G.; Memon, R.A. Comparative Analysis of Building Insulation Material Properties and Performance. *Renew. Sustain. Energy Rev.* **2020**, *131*, 110038. [CrossRef]
50. He, Y.; Zhang, Y.; Yao, Y.; He, Y.; Sheng, X. Review on the Prediction and Control of Structural Vibration and Noise in Buildings Caused by Rail Transit. *Buildings* **2023**, *13*, 2310. [CrossRef]
51. Chapain, S.; Aly, A.M. Vibration Attenuation in a High-Rise Hybrid-Timber Building: A Comparative Study. *Appl. Sci.* **2023**, *13*, 2230. [CrossRef]
52. Jang, E.-S. Sound Absorbing Properties of Selected Green Material—A Review. *Forests* **2023**, *14*, 1366. [CrossRef]
53. Kang, C.-W.; Jang, S.-S.; Hashitsume, K.; Kolya, H. Estimation of Impact Sound Reduction by Wood Flooring Installation in a Wooden Building in Korea. *J. Build. Eng.* **2023**, *64*, 105708. [CrossRef]
54. Butorina, M.; Kuklin, D.; Drozdova, L. Application of Soundproof Planning to Reduce Noise in Residential Building. *IOP Conf. Ser. Earth Environ. Sci.* **2021**, *666*, 042006. [CrossRef]
55. Sahana, S.; Karthigayini, S. Design Strategies to Reduce the Impact of Visual and Noise Pollution in Urban Areas. *Asian Rev. Environ. Earth Sci.* **2020**, *7*, 67–71. [CrossRef]
56. Ouakka, S.; Verlinden, O.; Kouroussis, G. Railway Ground Vibration and Mitigation Measures: Benchmarking of Best Practices. *Railw. Eng. Sci.* **2022**, *30*, 1–22. [CrossRef]
57. He, W.; Zou, C.; Pang, Y.; Wang, X. Environmental Noise and Vibration Characteristics of Rubber-Spring Floating Slab Track. *Environ. Sci. Pollut. Res.* **2021**, *28*, 13671–13689. [CrossRef] [PubMed]
58. Deng, S.; Ren, J.; Zhang, K.; Xiao, Y.; Liu, J.; Du, W.; Ye, W.-L. Performance of Floating Slab Track with Rubber Vibration Isolator. *Int. J. Struct. Stab. Dyn.* **2024**, *24*, 2450065. [CrossRef]
59. Malmborg, J.; Flodén, O.; Persson, P.; Persson, K. Numerical Study on Train-Induced Vibrations: A Comparison of Timber and Concrete Buildings. *Structures* **2024**, *62*, 106215. [CrossRef]
60. Ming, X.; Zheng, J.; Wang, L.; Zhao, C.; Wang, P.; Yao, L. A Case Study of Excessive Vibrations inside Buildings Due to an Underground Railway: Experimental Tests and Theoretical Analysis. *J. Cent. South. Univ.* **2022**, *29*, 313–330. [CrossRef]
61. López-Mendoza, D.; Connolly, D.P.; Romero, A.; Kouroussis, G.; Galvín, P. A Transfer Function Method to Predict Building Vibration and Its Application to Railway Defects. *Constr. Build. Mater.* **2020**, *232*, 117217. [CrossRef]
62. He, C.; Zhou, S.; Guo, P. An Efficient Three-Dimensional Method for the Prediction of Building Vibrations from Underground Railway Networks. *Soil Dyn. Earthq. Eng.* **2020**, *139*, 106269. [CrossRef]
63. Hao, Y.; Qi, H.; Liu, S.; Nian, V.; Zhang, Z. Study of Noise and Vibration Impacts to Buildings Due to Urban Rail Transit and Mitigation Measures. *Sustainability* **2022**, *14*, 3119. [CrossRef]
64. Zheng, J.; Li, X.; Zhang, X.; Bi, R.; Qiu, X. Structure-Borne Noise of Fully Enclosed Sound Barriers Composed of Engineered Cementitious Composites on High-Speed Railway Bridges. *Appl. Acoust.* **2022**, *192*, 108705. [CrossRef]
65. Sadeghi, J.; Vasheghani, M. Safety of Buildings against Train Induced Structure Borne Noise. *Build. Environ.* **2021**, *197*, 107784. [CrossRef]
66. Sadeghi, J.; Vasheghani, M.; Khajehdezfuly, A. Propagation of Structure-Borne Noise in Building Adjacent to Subway Lines. *Constr. Build. Mater.* **2023**, *401*, 132765. [CrossRef]
67. Nering, K. Perception of Structure-Borne Sound in Buildings in Context of Vibration Comfort of Human in the Buildings. *IOP Conf. Ser. Mater. Sci. Eng.* **2020**, *960*, 022033. [CrossRef]
68. Wang, S.; Su, D.; Zhu, S.; Zhang, Q. Comparative Life Cycle Assessment of Raised Flooring Products. In *Sustainable Product Development*; Springer International Publishing: Cham, Switzerland, 2020; pp. 293–310.
69. Kudryavtsev, M. *Experimental Evaluation of the Seismic Resistance of the Raised Floor Structure*; AIP Publishing: Long Island, NY, USA, 2023; p. 060007.
70. Wang, J.; Du, B. Sound Insulation Performance of Foam Rubber Damping Pad and Polyurethane Foam Board in Floating Floors. *Exp. Tech.* **2023**, *47*, 839–850. [CrossRef]
71. Scoczynski Ribeiro, R.; Henneberg, F.; Catai, R.; Arnela, M.; Avelar, M.; Amarilla, R.S.D.; Wille, V. Assessing Impact Sound Insulation in Floating Floors Assembled from Construction and Demolition Waste. *Constr. Build. Mater.* **2024**, *416*, 135196. [CrossRef]
72. Crispin, C.; Dijckmans, A.; Wuyts, D.; Prato, A.; Rizza, P.; Schiavi, A. Analytical Prediction of Floating Floors Impact Sound Insulation Including Thickness-Resonance Wave Effects. *Appl. Acoust.* **2023**, *208*, 109380. [CrossRef]
73. Schoenwald, S.; Vallely, S. Cross Laminated Timber Elements with Functional Grading and Localised Ballast to Improve Airborne and Impact Sound Insulation. In Proceedings of the 10th Convention of the European Acoustics Association Forum Acusticum 2023, Turin, Italy, 17 January 2024; European Acoustics Association: Turin, Italy; pp. 1529–1534.
74. Nilsson, E.; Ménard, S.; Bard, D.; Hagberg, K. Effects of Building Height on the Sound Transmission in Cross-Laminated Timber Buildings—Airborne Sound Insulation. *Build. Environ.* **2023**, *229*, 109985. [CrossRef]

75. Mateus, D.; Pereira, A. The Strong Influence of Small Construction Errors on Sound Insulation of Buildings with Heavyweight Concrete Structure: A Portuguese Case Study. *Build. Acoust.* **2023**, *30*, 227–246. [CrossRef]
76. Fedorczak-Cisak, M.; Kowalska-Koczwara, A.; Nering, K.; Pachla, F.; Radziszewska-Zielina, E.; Stecz, P.; Tatara, T.; Jeleński, T. Measurement and Diagnosis of Comfort in a Historic Building. *Energies* **2022**, *15*, 8963. [CrossRef]
77. Papadakis, N.; Katsaprakakis, D.A.l. A Review of Energy Efficiency Interventions in Public Buildings. *Energies* **2023**, *16*, 6329. [CrossRef]
78. Amanowicz, Ł.; Ratajczak, K.; Dudkiewicz, E. Recent Advancements in Ventilation Systems Used to Decrease Energy Consumption in Buildings—Literature Review. *Energies* **2023**, *16*, 1853. [CrossRef]
79. Tognon, G.; Marigo, M.; De Carli, M.; Zarrella, A. Mechanical, Natural and Hybrid Ventilation Systems in Different Building Types: Energy and Indoor Air Quality Analysis. *J. Build. Eng.* **2023**, *76*, 107060. [CrossRef]
80. Abramkina, D. Noise from Mechanical Ventilation Systems in Residential Buildings. In *E3S Web of Conferences*; EDP Sciences: Les Ulis, France, 2023; Volume 457, p. 02007. [CrossRef]
81. Zekic, S.; Gomez-Agustina, L.; Aygun, H.; Chaer, I. Experimental Acoustic Testing of Alternative Ventilation Ducts. In Proceedings of the INTER-NOISE and NOISE-CON Congress and Conference Proceedings, Grand Rapids, MI, USA, 15–18 May 2023; Volume 265, pp. 2785–2792. [CrossRef]
82. Nering, K.; Kowalska-Koczwara, A.; Shymanska, A.; Pawluś, M. The Possibility of Providing Acoustic Comfort in Hotel Rooms as an Element of Sustainable Development. *Sustainability* **2022**, *14*, 13692. [CrossRef]
83. Nalls, L.; Daroux, P. Case Study: Effect of Mechanical Short-Circuiting of Isolation Pads on Rooftop Equipment Noise and Vibration Transfer to Spaces Below. In Proceedings of the INTER-NOISE and NOISE-CON Congress and Conference Proceedings, Grand Rapids, MI, USA, 15–18 May 2023; Volume 266, pp. 234–242. [CrossRef]
84. Mak, C.M.; Ma, K.W.; Wong, H.M. *Prediction and Control of Noise and Vibration from Ventilation Systems*; CRC Press: Boca Raton, FL, USA, 2023; ISBN 9781003201168.
85. Ruggieri, S.; Vukobratović, V. The Influence of Torsion on Acceleration Demands in Low-Rise RC Buildings. *Bull. Earthq. Eng.* **2024**, *22*, 2433–2468. [CrossRef]
86. Pečnik, J.G.; Gavrić, I.; Sebera, V.; Kržan, M.; Kwiecień, A.; Zając, B.; Azinović, B. Mechanical Performance of Timber Connections Made of Thick Flexible Polyurethane Adhesives. *Eng. Struct.* **2021**, *247*, 113125. [CrossRef]
87. Rousakis, T.; Papadouli, E.; Sapalidis, A.; Vanian, V.; Ilki, A.; Halici, O.F.; Kwiecień, A.; Zając, B.; Hojdys, Ł.; Krajewski, P.; et al. Flexible Joints between RC Frames and Masonry Infill for Improved Seismic Performance—Shake Table Tests. In *Brick and Block Masonry—From Historical to Sustainable Masonry*; CRC Press: Boca Raton, FL, USA, 2020; pp. 499–507.
88. Pessiki, S. Sustainable Seismic Design. *Procedia Eng.* **2017**, *171*, 33–39. [CrossRef]
89. Chaudhari, D.J.; Dhoot, G.O. Performance Based Seismic Design of Reinforced Concrete Building. *Open J. Civil Eng.* **2016**, *06*, 188–194. [CrossRef]
90. Sullivan, T.J.; Welch, D.P.; Calvi, G.M. Simplified Seismic Performance Assessment and Implications for Seismic Design. *Earthq. Eng. Eng. Vib.* **2014**, *13*, 95–122. [CrossRef]
91. Zhang, L. Vibration and Noise Reduction for Building Floor Structures Based on New Composite Material Vibration Isolation Pads. *Mater. Express* **2023**, *13*, 1944–1953. [CrossRef]
92. Luo, W.; Liang, Q.; Zhou, Y.; Lu, Z.; Li, J.; He, Z. A Novel Three-Dimensional Isolation Bearing for Buildings Subject to Metro- and Earthquake-Induced Vibrations: Laboratory Test and Application. *J. Build. Eng.* **2024**, *86*, 108798. [CrossRef]
93. Wu, D.; Tesfamariam, S.; Xiong, Y. FRP-Laminated Rubber Isolator: Theoretical Study and Shake Table Test on Isolated Building. *J. Earthq. Eng.* **2023**, *27*, 1302–1323. [CrossRef]
94. Li, Z.; Ma, M.; Liu, K.; Jiang, B. Performance of Rubber-Concrete Composite Periodic Barriers Applied in Attenuating Ground Vibrations Induced by Metro Trains. *Eng. Struct.* **2023**, *285*, 116027. [CrossRef]
95. Wang, T.; Jiang, B.; Sun, X. Train-Induced Vibration Prediction and Control of a Metro Depot and Over-Track Buildings. *Buildings* **2023**, *13*, 1995. [CrossRef]
96. Chiaro, G.; Palermo, A.; Banasiak, L.; Tasalloti, A.; Granello, G.; Hernandez, E. Seismic Response of Low-Rise Buildings with Eco-Rubber Geotechnical Seismic Isolation (ERGSI) Foundation System: Numerical Investigation. *Bull. Earthq. Eng.* **2023**, *21*, 3797–3821. [CrossRef]
97. Major, M.; Adamczyk, I.; Kalinowski, J. An Innovative Absorption Propagation System Hollow Block Made of Concrete Modified with Styrene–Butadiene Rubber and Polyethylene Terephthalate Flakes to Reduce the Propagation of Mechanical Vibrations in Walls. *Materials* **2023**, *16*, 5028. [CrossRef]
98. Abate, G.; Fiamingo, A.; Massimino, M.R.; Pitilakis, D.; Anastasiadis, A.; Vratsikidis, A.; Kapouniaris, A. Gravel-Rubber Mixtures as Geotechnical Seismic Isolation Systems underneath Structures: Large-Scale Tests vs FEM Modelling. In *Geosynthetics: Leading the Way to a Resilient Planet*; CRC Press: London, UK, 2023; pp. 1170–1176.
99. Zhao, Z.; Jiang, H.; Li, X.; Su, X.; Wu, X.; Zhang, X.; Zou, M. Study on the Effect of Different Polysiloxane Surfactants on the Properties of Polyurethane Microcellular Eelastomers. In Proceedings of the Ninth International Conference on Energy Materials and Electrical Engineering (ICEMEE 2023), Kuala Lumpur, Malaysia, 8–10 June 2023; Zhou, J., Aris, I.B., Eds.; SPIE: Pontoise, France, 2024; p. 151.
100. Nering, K.; Kowalska-Koczwara, A. Determination of Vibroacoustic Parameters of Polyurethane Mats for Residential Building Purposes. *Polymers* **2022**, *14*, 314. [CrossRef]

101. He, H.; Liu, Y.; Li, Y.; Shi, D.; Chen, Y.; Fan, H. Sandwich Meta-Panels for Vibration and Explosion Attenuation: Manufacturing, Testing, and Analyzing. *Int. J. Impact Eng.* **2023**, *177*, 104588. [CrossRef]
102. Huang, X.; Zeng, Z.; Li, Z.; Luo, X.; Yin, H.; Wang, W. Experimental Study on Vibration Characteristics of the Floating Slab with Under-Slab Polyurethane Mats Considering Fatigue Loading Effect. *Eng. Struct.* **2023**, *276*, 115322. [CrossRef]
103. Frescura, A.; Lee, P.J.; Schöpfer, F.; Schanda, U. Correlations between Standardised and Real Impact Sound Sources in Lightweight Wooden Structures. *Appl. Acoust.* **2021**, *173*, 107690. [CrossRef]
104. Nurzyński, J.; Nowotny, Ł. Acoustic Performance of Floors Made of Composite Panels. *Materials* **2023**, *16*, 2128. [CrossRef] [PubMed]
105. Nowotny, Ł.; Nurzyński, J. The Influence of Insulating Layers on the Acoustic Performance of Lightweight Frame Floors Intended for Use in Residential Buildings. *Energies* **2020**, *13*, 1217. [CrossRef]
106. Kim, J.-H.; Park, H.-G.; Han, H.-K.; Mun, D.-H. Dynamic Properties of Floating Floor System and Its Influence on Heavy-Weight Impact Sound in Residential Buildings. *J. Acoust. Soc. Am.* **2021**, *150*, 1251–1263. [CrossRef] [PubMed]
107. Adžić, S.; Radičević, B.; Bjelić, M.; Šimunović, D.; Marašević, M.; Stojić, N. Measuring the Insulating Power from the Sound Impact of the Floor Covering of Pressed Expanded Polystyrene. *J. Prod. Eng.* **2023**, *26*, 1–6. [CrossRef]
108. Xu, G.; Guo, T.; Li, A.; Zhang, H.; Wang, K.; Xu, J.; Dang, L. Seismic Resilience Enhancement for Building Structures: A Comprehensive Review and Outlook. *Structures* **2024**, *59*, 105738. [CrossRef]
109. Hu, X.; Zhao, Z.; Barredo, E.; Dai, K.; Tang, Y.; Hong, N. Nonlinear Viscous Damping-Based Negative Stiffness Isolation System for Over-Track Complex Structures. *J. Earthq. Eng.* **2024**, 1–29. [CrossRef]
110. Wang, Y.; He, Z.; Wang, K.; Li, P.; Yun, J. Research on the Damping Performance of a Sleeper Damping Track with Elastic Side-Supporting Pad. *Constr. Build. Mater.* **2024**, *420*, 135648. [CrossRef]
111. Cruz, C.; Miranda, E. Damping Ratios of the First Mode for the Seismic Analysis of Buildings. *J. Struct. Eng.* **2021**, *147*, 04020300. [CrossRef]
112. Tulebekova, S.; Malo, K.A.; Rønnquist, A.; Nåvik, P. Modeling Stiffness of Connections and Non-Structural Elements for Dynamic Response of Taller Glulam Timber Frame Buildings. *Eng. Struct.* **2022**, *261*, 114209. [CrossRef]
113. Mantakas, A.G.; Kapasakalis, K.A.; Alvertos, A.E.; Antoniadis, I.A.; Sapountzakis, E.J. A Negative Stiffness Dynamic Base Absorber for Seismic Retrofitting of Residential Buildings. *Struct. Control Health Monit.* **2022**, *29*. [CrossRef]
114. Giresini, L.; Taddei, F.; Solarino, F.; Mueller, G.; Croce, P. Influence of Stiffness and Damping Parameters of Passive Seismic Control Devices in One-Sided Rocking of Masonry Walls. *J. Struct. Eng.* **2022**, *148*, 04021257. [CrossRef]
115. Nering, K.; Nering, K. A Low-Stress Method for Determining Static and Dynamic Material Parameters for Vibration Isolation with the Use of VMQ Silicone. *Materials* **2023**, *16*, 2960. [CrossRef]
116. Nering, K. Damping Problem of Numerical Simulation of Structural Reverberation Time. *IOP Conf. Ser. Mater. Sci. Eng.* **2020**, *960*, 042083. [CrossRef]
117. Papagiannopoulos, G.A.; Hatzigeorgiou, G.D. On the Use of the Half-Power Bandwidth Method to Estimate Damping in Building Structures. *Soil Dyn. Earthq. Eng.* **2011**, *31*, 1075–1079. [CrossRef]
118. Wu, B. A Correction of the Half-Power Bandwidth Method for Estimating Damping. *Arch. Appl. Mech.* **2015**, *85*, 315–320. [CrossRef]
119. Yu, D.-H.; Li, G.; Li, H.-N. Implementation of Rayleigh Damping for Local Nonlinear Dynamic Analysis Based on a Matrix Perturbation Approach. *J. Struct. Eng.* **2021**, *147*, 04021130. [CrossRef]
120. Yılmaz, İ.; Arslan, E.; Kızıltaş, E.Ç.; Çavdar, K. Development of a Prediction Method of Rayleigh Damping Coefficients for Free Layer Damping Coatings through Machine Learning Algorithms. *Int. J. Mech. Sci.* **2020**, *166*, 105237. [CrossRef]
121. Sánchez Iglesias, F.; Fernández López, A. Rayleigh Damping Parameters Estimation Using Hammer Impact Tests. *Mech. Syst. Signal Process* **2020**, *135*, 106391. [CrossRef]
122. Wu, H.-Y.; Rubinstein, M.; Shih, E.; Guttag, J.; Durand, F.; Freeman, W. Eulerian Video Magnification for Revealing Subtle Changes in the World. *ACM Trans. Graph.* **2012**, *31*, 65. [CrossRef]
123. Qiu, Q. Automated Defect Detection in FRP-Bonded Structures by Eulerian Video Magnification and Adaptive Background Mixture Model. *Autom. Constr.* **2020**, *116*, 103244. [CrossRef]

materials

Article

Experimental Study on the Microfabrication and Mechanical Properties of Freeze–Thaw Fractured Sandstone under Cyclic Loading and Unloading Effects

Taoying Liu [1,2,*], **Wenbin Cai** [1], **Yeshan Sheng** [1] **and Jun Huang** [1,2]

1   School of Resource and Safety Engineering, Central South University, Changsha 410083, China
2   Guangxi Liubin Expressway Construction and Development Co., Ltd., Nanning 530023, China
*   Correspondence: taoying@csu.edu.cn; Tel.: +86-15111120528

Abstract: A series of freeze–thaw cycling tests, as well as cyclic loading and unloading tests, have been conducted on nodular sandstones to investigate the effect of fatigue loading and freeze–thaw cycling on the damage evolution of fractured sandstones based on damage mechanics theory, the microstructure and sandstone pore fractal theory. The results show that the number of freeze–thaw cycles, the cyclic loading level, the pore distribution and the complex program are important factors affecting the damage evolution of rocks. As the number of freeze–thaw cycles rises, the peak strength, modulus of elasticity, modulus of deformation and damping ratio of the sandstone all declined. Additionally, the modulus of elasticity and deformation increase nonlinearly as the cyclic load level rises. With the rate of increase decreasing, while the dissipation energy due to hysteresis increases gradually and at an increasing rate, and the damping ratio as a whole shows a gradual decrease, with a tendency to increase at a later stage. The NRM (Nuclear Magnetic Resonance) demonstrated that the total porosity and micro-pores of the sandstone increased linearly with the number of freeze–thaw cycles and that the micro-porosity was more sensitive to freeze–thaw, gradually shifting towards meso-pores and macro-pores; simultaneously, the SEM (Scanning Electron Microscope) indicated that the more freeze–thaw cycles there are, the more micro-fractures and holes grow and penetrate each other and the more loose the structure is, with an overall nest-like appearance. To explore the mechanical behavior and mechanism of cracked rock in high-altitude and alpine areas, a damage model under the coupling of freeze–thaw-fatigue loading was established based on the loading and unloading response ratio theory and strain equivalence principle.

Keywords: freeze–thaw cycles; rock mechanics; cyclic loading and unloading; microstructure

1. Introduction

It is estimated that nearly 50% of the world's seasonally frozen and permafrost areas are located around the globe [1]. With the rapid development of construction projects in cold regions, rock damage caused by freezing and thawing affects the durability and longevity of geostructures, especially those constructed in areas frequently affected by climate change [2,3]. Furthermore, excessively low temperatures for engineering building cause significant freeze–thaw disaster difficulties; the future development of a large number of cold locations will encounter ultra-low temperature frozen rock adverse geological environments [4]. Under the action of a freeze–thaw cycle, the permeability of the fractured rock mass is reduced, and water penetrates into it. After freezing, water ice phase change and volume expansion make the volume of water expand by about 9%, resulting in frost heave pressure and aggravating the development, initiation, convergence, expansion and penetration of rock mass cracks [5,6], which are often used as a cyclic or fatigue load, and their mechanical properties are very different from those under static load, such as tunnel blasting excavation and support processes, deep oil storage tank liquid circulation repeated input and output and the excavation and support of high slopes. The mining

Citation: Liu, T.; Cai, W.; Sheng, Y.; Huang, J. Experimental Study on the Microfabrication and Mechanical Properties of Freeze–Thaw Fractured Sandstone under Cyclic Loading and Unloading Effects. *Materials* **2024**, *17*, 2451. https://doi.org/10.3390/ma17102451

Academic Editor: René de Borst

Received: 11 April 2024
Revised: 13 May 2024
Accepted: 14 May 2024
Published: 19 May 2024

face is carried out alternately, which involves cyclic and repeated fatigue stresses. The deformation strength characteristics and fracture damage characteristics of rock are closely related to the stress path and loading history. Meanwhile, the macro and micro crack development and the expansion evolution of fractured rock mass are greatly affected by periodic loads.

At present, with the continuous development of science and technology, numerous scholars have observed the microscopic damage of rocks after freeze–thaw cycles with the help of advanced equipment, such as NMR, scanning electron microscopes, CT, rock wave velocity meters, morphological scanners, etc. Niu Caoyan et al. [7] investigated the effect of freeze–thaw cycling on the porosity and fracture surface microstructure of rocks with the aid of nuclear magnetic resonance (NMR) and scanning electron microscopy (SEM), showing that the higher the number of freeze–thaw cycles, the lower the porosity of the rock. Li Jielin [8–10] classified the internal pores of rocks into small, medium and large pores according to the pore size distribution characteristics, and it was obvious that small and medium pores kept developing and large pores kept expanding under the effect of freezing and swelling pressure, and it was also found that the NMR porosity and T2 spectral area of rocks showed exponentially decreasing distribution characteristics with the uniaxial compressive strength of rocks. X-ray CT scans are mostly utilized in medicine and industries. It is a method of inspecting the interior deterioration of materials. Benjamin K. Blykers scanned mineral building materials using X-ray dark-field imaging (DFI) to analyze the micropores inside the materials through dark-field signals [11]. CT technology is increasingly being utilized to examine the interior damage features of rocks. Koorosh Abdolghanizadeh et al. [12], Wang Y et al. [13] and Liu Hui et al. [14] compared the damage characteristics of rocks under different freeze–thaw cycles by the CT technique, and the study showed that with the higher number of freeze–thaw cycles, the CT value increased, the internal crack development of rocks kept expanding, the structure was obviously defective and deteriorated and the internal damage was boosted.

The deep rock is not only in the state of high ground stress but is also affected by stress disturbances such as mining, drilling and blasting, which cause stress redistribution, internal structure changes and even engineering instability. There are various types of cyclic loading and unloading, including equal amplitude cyclic loading and unloading, graded cyclic loading and unloading, variable lower limit cyclic loading and unloading and constant lower limit cyclic loading and unloading. Liu Xiangyu et al. [15] carried out equal-amplitude cyclic and step-by-step primary cyclic unloading tests on siltstone to compare the pore expansion, crack production and crack evolution patterns in the damage phase under the two unloading methods. The test results show that the porosity of the two types of unloaded rocks tends to increase first and then decrease, and the rock samples are mainly damaged by shear; the damage to the equal-amplitude cyclic unloading rock samples occurs during the loading stage, displaying instantaneous damage characteristics, whereas the damage to the rock samples arises during the unloading stage, displaying small amplitude oscillation damage features. Wang Tianzuo et al. [16] applied acoustic emission to compare the mechanical properties of constant lower limit cyclic loading and unloading and variable lower limit cyclic loading and unloading under uniaxial compression tests and found that constant lower limit cyclic loading and unloading increased the rock strength by 6.5%. The complexity of fatigue damage is so significant that the evolution of the damage is difficult to describe in terms of general mechanical theory. As the deformation and damage of rocks are essentially caused by energy release, energy analysis is an important and critical method for revealing the evolution of damage in rocks under disturbed stress conditions. To determine the total rock energy, elastic strain energy and dissipation energy at various fracture dips, Peng Kang et al. [17] performed stress gradient continuous cycle loading and unloading studies on sandstones with varied fracture dips. Ju Wang et al. [18,19] investigated the characteristics of energy storage and dissipation, damage evolution and failure modes of red sandstone under discontinuous multistage fatigue (DMLF) loading at different fracture dips, and the results showed that the total energy, elastic energy and dissipation energy showed a quadratic polynomial function increase with increasing stress.

In summary, NMR, SEM and CT can effectively observe the microscopic damage of rocks after freeze–thaw cycles. In the meantime, there is a lack of research on the damage evolution of fractured rock under freeze–thaw-fatigue unloading loading coupling effects. Consequently, based on the results of previous studies [20,21], NMR and SEM are used to investigate rock damage under coupled freeze–thaw-fatigue loading as a means of enhancing the rigor of the results to complement the gaps in previous studies. It is not only an important research frontier in the discipline but also a significant social and engineering background and is of great practical significance in predicting the occurrence of geoengineering hazards in alpine and high-altitude areas affected by disturbed loading.

2. Experimental System and Methods

This experiment was designed according to the Code for rock tests in water and hydropower projects (SL/T 264-2020). A yellow sandstone from a mine was selected and machined into a 50 mm × 50 mm × 100 mm square specimen with a 10 mm long, 2 mm wide, 45° inclined prefabricated fracture through the middle. In this experiment, three sandstone specimens were tested at each set of freeze–thaw cycles, and the statistical data obtained from the experiment were averaged to minimize experimental errors [22]. The basic mechanical parameters of sandstone are shown in Table 1, and the mineralogical composition of the sandstone is shown in Figure 1. The sandstone was saturated with water before the test began and then placed in a TDS-300 freeze–thaw tester to start the freeze–thaw damage test. An AiniMR-150 NMR (Shanghai Newmy Electronics Technology Co., Ltd., Shanghai, China) instrument was used to measure the porosity changes before and after the freeze–thaw cycle and an HS-YS4A rock wave velocity meter (Beijing Haifuda Technology Co., Ltd., Beijing, China) was used to determine the wave velocity of sandstone before and after freeze–thaw damage, followed by a WHY-300/10 microcomputer-controlled electro-hydraulic servo universal testing machine (Shanghai Hualong Test Instruments Co., Ltd., Shanghai, China) carrying out the cycle plus unloading test and then removing the damaged fragments for SEM microstructure observation. The test process is shown in Figure 2.

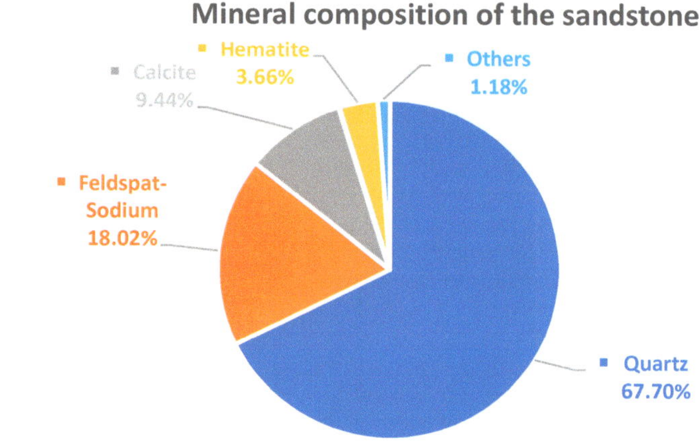

Figure 1. Mineral composition of the sandstone.

Table 1. Basic Mechanical Parameters of Sandstone.

Compressive Strength (MPa)	Modulus of Elasticity (GPa)	Tensile Strength (MPa)	Poisson's Ratio ν
44.72	3.72	2.57	0.30

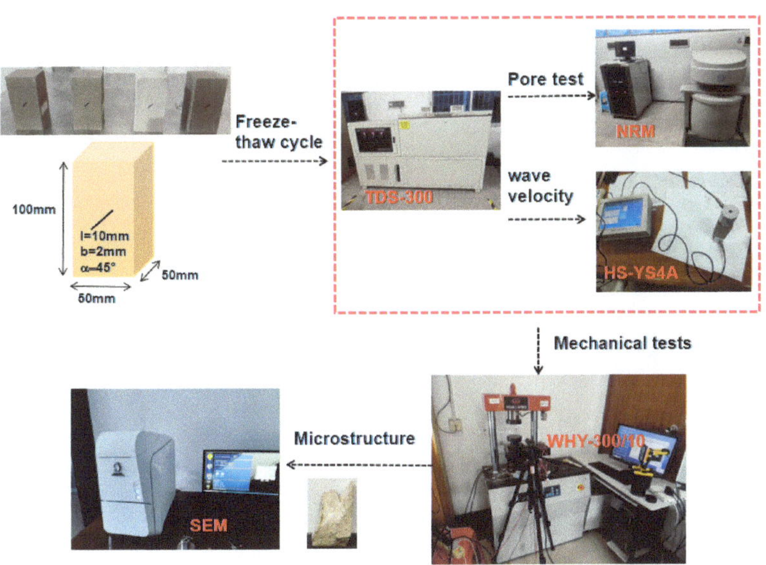

Figure 2. Test systems.

The freeze–thaw cycle was executed in five sets of 0, 20, 40, 60 and 80 freeze–thaw cycles, with three specimens made under each test condition. For each cycle, the freezing time is 240 min, the melting time is 240 min and the freezing and melting temperatures are −20 °C and 20 °C, respectively. It takes about 1.5 h for each cycle to decrease from a normal temperature of 20 °C to −20 °C and then from a freezing temperature to melting. The cumulative duration of a freeze–thaw cycle is about 11 h, as shown in Figure 3a. The fatigue test was performed in a constant down-line cyclic loading and unloading mode with force control and a loading rate of 0.5 kN/s; the loading mode was 0 kN→11 kN→1 kN→21 kN→1 kN→31 kN→1 kN..., as indicated in Figure 3b.

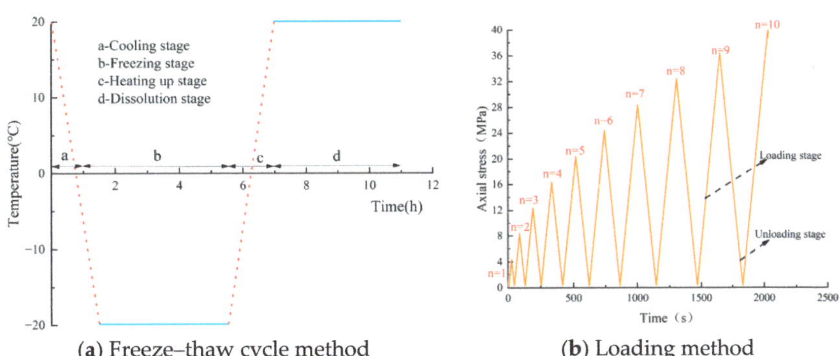

(**a**) Freeze–thaw cycle method   (**b**) Loading method

Figure 3. Test method.

3. Analysis of Experimental Results

3.1. Stress–Strain Curves

Figure 4 shows the cyclic stress–strain curves for the fractured sandstone at 0, 20, 40, 60 and 80 freeze–thaw cycles. The hysteresis curve spacing of the specimens shows an overall "dense and sparse" pattern, and the higher the number of freeze–thaw cycles, the sparser the overall hysteresis curve spacing. In the early stages of loading, the peak value of each stage of loading was almost on the same curve while gradually shifting in the direction of increasing strain, and this curve was found by most scholars to coincide with the uniaxial compression curve of the specimen, while in the late stages of loading, the curve deviated significantly from the historical stress–strain curve [23,24], which indicates that excessive fatigue loading in the early stages led to the accumulation of damage inside the specimen and the gradual weakening of the bearing capacity.

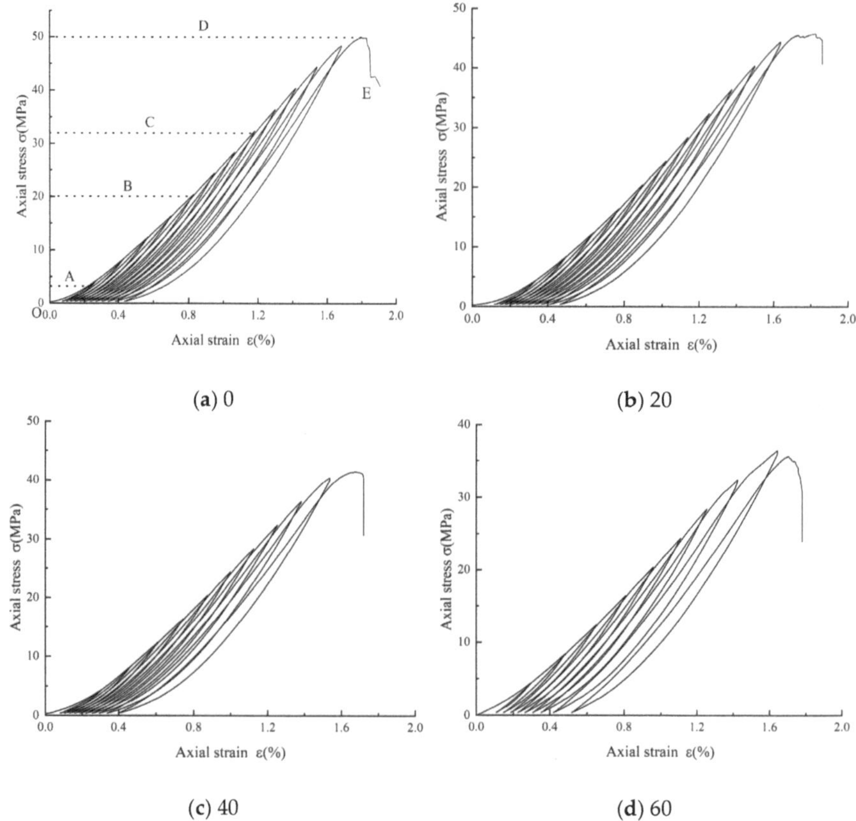

Figure 4. *Cont.*

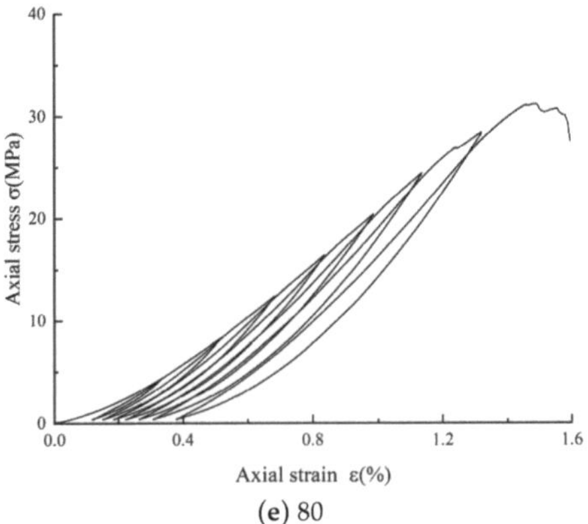

(e) 80

Figure 4. Stress–strain curve.

The evolution of the stress–strain curve throughout the cyclic loading and unloading test of the post-freeze–thaw fractured rock consists of five stages, as shown in Figure 4a. There are five stages in the evolution of the stress–strain curve during the cyclic loading and unloading test of the post-freeze–thaw fractured rock [25].

(1) The OA compacting stage: This stage is the initial loading stage of the specimen, with the increase in axial stress, the specimen internal microcracks and pore compacting. The curve shows non-linear growth, and it was also found that with the increase in the number of freeze–thaw cycles, the faster the change in axial stress, indicating that there is freeze–thaw action due to the water in the rock micro-porosity. The water–ice phase change produces about 9% volume expansion, making the micro-porosity further increase.

(2) The AB elastic deformation stage: This stage keeps the curve approximately straight up as the axial force continues to increase. This stage is elastic deformation and can be recovered.

(3) The BC crack stable expansion stage: As the axial force increases, the curve exhibits a non-linear growth trend and the microcracks within the specimen begin to expand with a gradual increase in density and in the direction of the maximum principal stress.

(4) The CD crack instability expansion stage: In this stage, with the further increase in the axial force, cracks occur and gradually gather into nuclei, the micro-crack expansion rate increases rapidly and eventually reaches the peak strength and the specimen is damaged.

(5) The DE damage stage: The deformation of the sandstone increases rapidly under the continuous action of axial stress, and after reaching the compressive strength of the specimen, the load-bearing capacity decreases rapidly and the stress–strain curve falls off rapidly. There is obvious brittle damage of the specimen, which maintains a certain residual strength due to the presence of shear strength and friction on the fracture surface of the specimen.

3.2. Mechanical Parameters

The mass and wave velocity of sandstone before and after freezing and thawing were measured, and the damage of the mass and wave velocity was calculated, as shown in Equations (1) and (2), where D_m is the mass damage factor, $\overline{m_0}$ and $\overline{m_i}$ are the average mass (kg) of the sandstone before and after the freeze–thaw cycle, respectively, D_v is the wave velocity damage factors and V_i is the mean wave velocities (m/s) of the sandstone before and after the freeze–thaw cycle, respectively.

$$D_{\mathrm{m}} = \frac{\overline{m_0} - \overline{m_i}}{\overline{m_0}} \tag{1}$$

$$D_{\mathrm{v}} = \frac{\overline{V_0} - \overline{V_i}}{\overline{V_0}} \tag{2}$$

Figure 5 represents the relationship between the mass and wave damage factor after freeze–thaw cycles in sandstone and the number of freeze–thaw cycles. As the number of freeze–thaw cycles increases, the overall mass and wave velocity damage factors show a pattern of increasing. The low number of freeze–thaw cycles has a small effect on sandstone mass loss, while the high number of freeze–thaw cycles leads to a sudden increase in the mass damage factor and more mass loss, indicating that dislocation and movement between mineral grains within the sandstone under the action of water–ice phase change can lead to particle loss and rock chips falling off the surface. The wave speed is a reflection of the extent of internal defects (microcracks, pores and joints, etc.) in the rock; the lower the wave speed, the more serious the internal defects [26]. It is not difficult to find that with the increase in the number of freeze–thaw cycles, the wave velocity damage factor rises almost linearly, and the internal deterioration degree of sandstone is deepened. Moreover, it is found in the test that cracks appear on the prefabricated crack surface of sandstone after freeze–thaw damage, as well as the phenomenon of a frozen crisp at the edge.

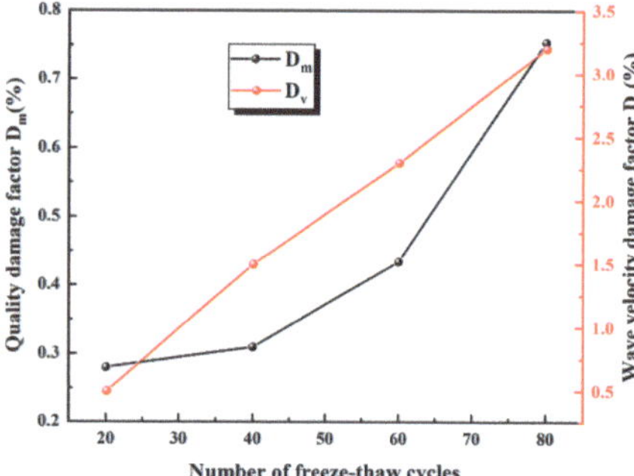

Figure 5. Damage factor.

The damage strength and damage rating of each sandstone are shown in Table 2. The sandstone with no freeze–thaw damage reaches a load cycle rating of 12, and as the number of freeze–thaw cycles gradually increases, the damage rating gradually decreases, culminating in a sandstone damage rating of 7 after 80 freeze–thaw cycles, a reduction of 5 grades compared to that with no freeze–thaw. As can be seen in Figure 6, the peak strength of the sandstone decreased almost linearly from 49.95 MPa to 31.17 MPa, a decrease of 37.6%, indicating that freeze–thawing has a significant effect on the deterioration of the sandstone bearing capacity. To further evaluate the degree of strength loss before and after freeze–thaw cycles, the freeze–thaw coefficient is often used. The larger the value, the less freeze–thaw damage and the better the resistance to freezing, and vice versa. As the number of freeze–thaw cycles increases, the freezing resistance of the sandstone decreases linearly from 0.911 to 0.624, a decrease of 31.5%, indicating that repeated freeze–thawing exacerbates the evolution of the sandstone damage and weakens the internal cementation between the grains.

$$K_f = \frac{\overline{R_f}}{\overline{R_s}} \tag{3}$$

where K_f denotes the frost resistance factor and $\overline{R_f}$ and $\overline{R_s}$ denote the peak strength after freeze–thaw cycles and the peak strength before freeze–thaw cycles, respectively.

$$\varepsilon_s = \varepsilon_e + \varepsilon_{cr} \tag{4}$$

where ε_s is the total strain in each phase, ε_e is the elastic strain in each phase and ε_{cr} is the plastic strain in each stage.

$$E_e = \frac{\sigma_{max} - \sigma_{min}}{\varepsilon_e} \tag{5}$$

$$E_d = \frac{\sigma_{max} - \sigma_{min}}{\varepsilon_s} \tag{6}$$

where E_e is the modulus of elasticity for each phase, E_d is the modulus of deformation in each stage and σ_{max} and σ_{min} are the maximum and minimum stresses in each stage, respectively.

$$\lambda = \frac{A_R}{4\pi A_s} \tag{7}$$

Table 2. Peak strength and breakage level of specimens.

Number of Freeze–Thaw Cycles	Specimen Number	Peak Strength σ_p (MPa)	Average Peak Strength σ_p (MPa)	Breakage Level
0	1	48.98	49.95	12
	2	49.20		
	3	51.59		
20	1	44.99	45.48	11
	2	45.25		
	3	46.2		
40	1	41.40	41.43	10
	2	40.99		
	3	41.9		
60	1	36.2	36.4	9
	2	36.7		
	3	36.3		
80	1	30.88	31.17	7
	2	31.7		
	3	30.93		

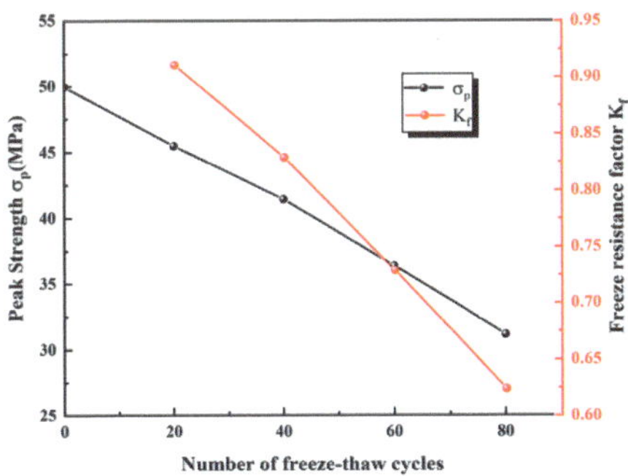

Figure 6. Relationship between the peak intensity and the number of freeze–thaw cycles.

Figure 7 represents the stress–strain hysteresis curve and calculates the modulus of elasticity, modulus of deformation and damping ratio, respectively. λ is the damping ratio, A_R is the area of hysteresis circle ABCD (kJ/m^3) and A_s is the area of the triangle AEF (kJ/m^3).

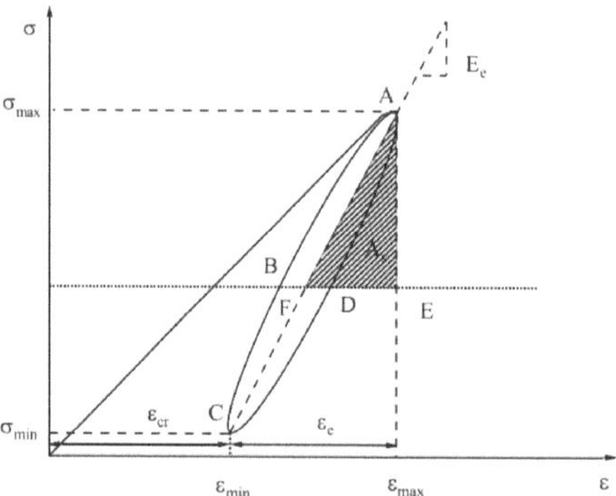

Figure 7. Schematic diagram of the hysteresis loop of the stress–strain curve.

Figure 8 represents the modulus of elasticity, modulus of deformation, damping ratio and area of hysteresis circle ABCD as a function of the number of freeze–thaw cycles. The modulus of elasticity reflects the ability of the sandstone to deform elastically, while the modulus of deformation reflects the deformation energy of the sandstone as a whole, including both elastic and plastic deformation [27,28]. It can be seen in Figure 8a,b that the modulus of elasticity and modulus of deformation gradually increase with increasing cyclic load levels at a constant number of freeze–thaw cycles, but the rate of increase is gradually slow, indicating that the load size increases step by step and the overall deformation increases. While the deformation at each level decreases, the load-bearing capacity of the sandstone is continuously weakened, internal microcracks expand and penetrate and damage gradually accumulates; at the same time, the deformation and resilience of the sandstone is continuously weakened and gradually turns into plastic damage. Additionally, the modulus of deformation increases abruptly when the cyclic load rating is increased from level 1 to level 2, mainly due to the compacting of the precast fractures and internal microcracks under the load. The higher the number of freeze–thaw cycles, the lower the modulus of elasticity and modulus of deformation of the sandstone. The main reason for this is that water enters the micropores of the sandstone, the water–ice phase change generates freezing pressure, followed by volume expansion, and repeated icing and melting increase the micro-pores, deteriorating the load-bearing capacity of the sandstone and leading to a significant reduction in the deformation capacity of the sandstone. It is also easy to see that sandstones with a high number of freeze–thaw cycles have a lower modulus of elasticity and modulus of deformation per load cycle class than sandstones with a low number of freeze–thaw cycles. This is because the greater the freeze–thaw damage to the sandstone under the same sized loading, the more the deformation and load-bearing capacity are weakened, resulting in the decay of the modulus of elasticity and modulus of deformation per level.

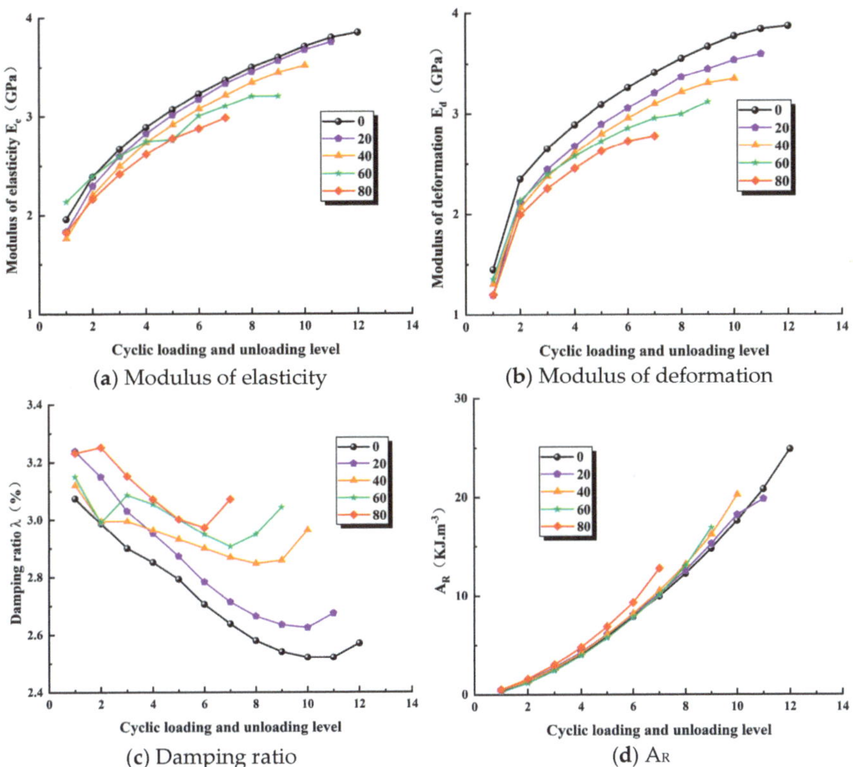

Figure 8. Mechanical parameters.

The damping ratio is a significant reflection of the mechanical properties of the rock under the cyclic loading of regional earthquakes. Under cyclic loading, the loading and unloading curves of sandstone do not coincide and exhibit hysteresis curves, i.e., reflecting the axial in damped vibration, while the area and shape of the hysteresis circle reflect the magnitude of dissipation energy generated by internal damage in the rock during a load cycle [29,30]. In Figure 8c, it can be seen that as the load cycle level increases, the damping ratio as a whole shows a pattern of gradually decreasing while having an increasing trend at a later stage. The author believes that as the load cycle level increases, the internal damage to the sandstone gradually accumulates, but the rate of damage development is slower, and when the sandstone damage load is reached, a large number of internal cracks in the sandstone expand, aggregate and penetrate, surface macro cracks are produced, and plastic deformation is more likely to occur, thus leading to a tendency for the damping ratio to increase at a later stage [31]. The higher the number of freeze–thaw cycles, the lower the damping ratio, and it fluctuates between 2.52% and 3.25%. This is mainly due to the fact that freeze–thaw aggravates the damage inside the sandstone and increases the porosity inside, resulting in greater plastic deformation of the sandstone and, hence, a greater damping ratio. As can be seen in Figure 8d, the dissipation energy due to hysteresis increases gradually and at an increasing rate as the cyclic load level increases. The difference in the magnitude of the dissipation energy release is initially small for sandstones subjected to different numbers of freeze–thaw cycles but gradually becomes larger in the later stages and increases abruptly under the final cyclic load. Similarly, more dissipative energy is released from sandstones subjected to a higher number of freeze–thaw cycles than from sandstones subjected to a lower number of freeze–thaw cycles at each load level. This is mainly due to the fact that the initial load is small, the number of internal microcracks is small and, therefore, the degree of damage is low, and the difference in the dissipation energy release is small for sandstones with different numbers of freeze–thaw cycles. This is when the dissipation energy increases dramatically. It is easy to explain that the higher the number of freeze–thaw cycles, the more porosity and micro-cracks are created in the sandstone and the higher the dissipation energy release per stage under the same load, while the sandstone undergoes fewer cycles and eventually the total dissipation energy release decreases due to the severe damage inside the sandstone exacerbated by freeze–thaw. The dissipative energy released from the sandstone for 0, 20, 40, 60 and 80 freeze–thaw cycles is 24.84 kJ/m^3, 19.78 kJ/m^3, 20.20 kJ/m^3, 16.84 kJ/m^3 and 12.70 kJ/m^3, respectively. The increase in the number of freeze–thaw cycles from 0 to 80 reduces the dissipative energy release by 12.14 kJ/m^3, a reduction of 48.87%; consequently, the freeze–thaw damage to sandstone is not negligible.

3.3. Microstructure

3.3.1. Nuclear Magnetic Resonance

To better reflect the microscopic pore changes in freeze–thaw sandstone, different numbers of freeze–thaw cycles were plotted based on T_2 spectra. The transverse relaxation rate of NMR (Nuclear Magnetic Resonance) can be expressed by the following equation according to NMR theory.

$$\frac{1}{T_2} = \rho_2 \frac{s}{v} = \frac{\rho_2 F_2}{r_c} \tag{8}$$

where T_2 is the lateral relaxation time (ms), ρ_2 is the surface relaxation strength, s is the pore surface area, v is the pore volume, F_2 is the core pore shape factor, usually a constant and pore shape dependent, and r_c is the pore radius of the rock. Let $\rho_2 F_2 = C$ and take $C = 10$; then, Equation (8) becomes Equation (9) [31].

$$r_c = CT_2 = 10T_2 \tag{9}$$

According to the sandstone T_2, the spectral peak curves are divided into different pore types according to different pore distributions, and the author, based on the experimental capillary pressure measurement porosity radius grading method and combined with relevant literature, divided the red sandstone pore size into three intervals, i.e., small pores ($r \leq 1$ μm), medium pores (1 μm $\leq r \leq 10$ μm) and large pores ($r \leq 10$ μm) [10,32,33]. Combined with Equation (9), it can be seen that the transverse relaxation time T_2 spectrum is distributed in the range of 0–10 ms for small pores, 10–100 ms for medium pores and above 100 ms for large pores or micro-fractures. The spectra showed a bimodal pattern, as shown in Figure 9a. As can be seen in Figure 9b, the cumulative porosity is almost zero within the relaxation time of 1 ms, while the cumulative porosity increases dramatically when the relaxation time increases from 1 ms to 10 ms, after which the cumulative porosity rises slowly and finally reaches the equilibrium value. It can be found that the cumulative porosity starts to differ slowly after the time of 10 ms for different numbers of freeze–thaw cycles, and the higher the number of freeze–thaw cycles, the lower the cumulative porosity, which indicates that the deterioration damage of the pore structure by freeze–thaw is gradually enhanced.

According to the above grading criteria, the different pore fractions of fracture dip 45° sandstone are counted and shown in Figure 10a. The total porosity and micro-porosity of the sandstone increased almost linearly as the number of freeze–thaw cycles increased from 0 to 80, the total porosity increased from 2.81% to 3.86%, an increase of 37.4%, and the microporosity increased from 2.22% to 2.93%, an increase of 32%, while the medium porosity increased to a lesser extent and the large porosity remained almost unchanged. The increase in mesoporosity was smaller, while the macroporosity remained almost unchanged and was less affected by freeze–thaw. The sandstone is dominated by micropores, which account for more than 70% of the total pores; see Figure 10b. Micro-pores develop rapidly under the influence of freeze–thaw, with the greatest increase in number, and are more sensitive to freeze–thaw, and the proportion of mesopores and macropores increases after 80 freeze–thaw cycles, indicating that the higher the number of freeze–thaw cycles, the sandstone gradually develops micropores into macropores, and the internal fine structure of the sandstone is gradually damaged.

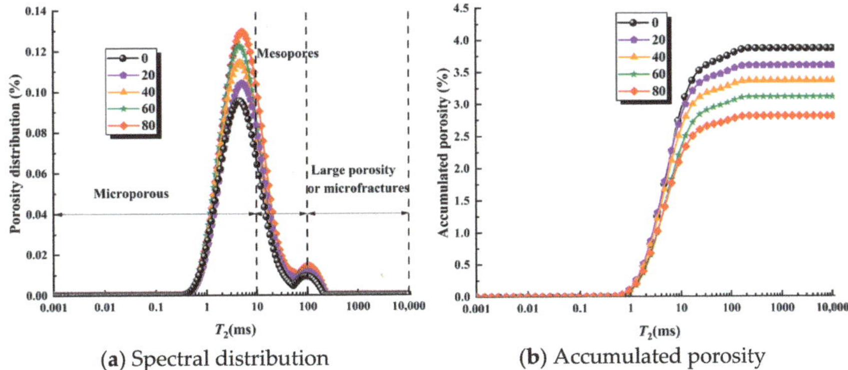

(**a**) Spectral distribution (**b**) Accumulated porosity

Figure 9. T_2 Spectral distribution and accumulated porosity.

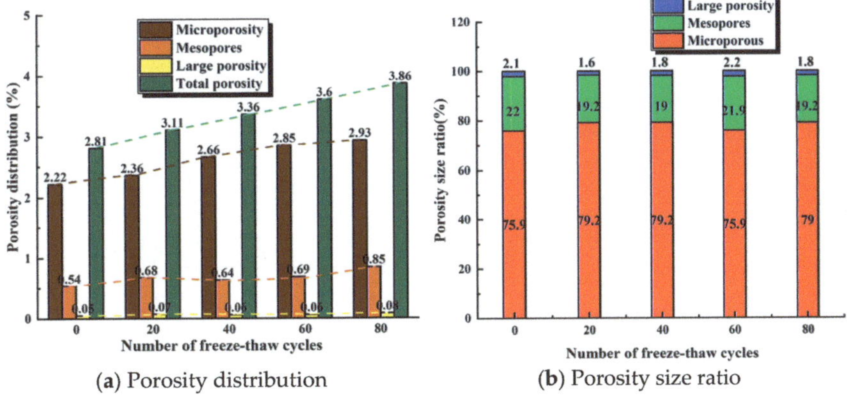

(**a**) Porosity distribution (**b**) Porosity size ratio

Figure 10. Changes in pore distribution with freeze–thaw cycles.

Since the porosity distribution is irregular and complex, the use of a simple geometric formulation to describe it is a complete failure to provide insight into porosity. However, it can be studied using fractal theory, i.e., it can be characterized using a fixed non-integer dimension between Euclidean dimensions [34]. The fractal dimension is a quantitative parameter that describes the degree of irregularity of a fractal object. The degree of irregularity of a fractal object, and thus the complexity and irregularity of the pore structure, can be indirectly reflected by this parameter.

According to the fractal theory [35], the number n of pores with diameters larger than *r* satisfies the following functional relationship.

$$n(>r) = \int_r^{r_{max}} I(r)dr = ar^{-D} \tag{10}$$

The volume of a pore with a pore size less than *r* is denoted as

$$V(<r) = \int_{r_{min}}^r I(r)ar^3dr \tag{11}$$

where r_{min} and r_{max} are the minimum and maximum porosity, respectively, $I(r)$ is the pore size distribution density, a is a constant and D is the pore fractal dimension.

Combining Equations (10) and (11) yields

$$V(<r) = \beta\left(r^{3-D} - r_{min}^{3-D}\right) \tag{12}$$

where β is a constant.

The cumulative pore volume fraction for pore sizes smaller than r is expressed as

$$S_V = \frac{V(<r)}{V_S} = \frac{r^{3-D} - r_{min}^{3-D}}{r_{max}^{3-D} - r^{3-D}} \tag{13}$$

where V_S is the total porosity.

Due to $r_{min} \ll r_{max}$, Equation (13) can be simplified as

$$S_V = \frac{r^{3-D}}{r_{max}^{3-D}} \tag{14}$$

According to Equation (9), T_2 is proportional to r. Consequently, Equation (14) can be written as

$$S_V = \left(\frac{T_2}{T_{2,max}}\right)^{3-D} \tag{15}$$

where $T_{2,max}$ is the maximum relaxation time.

Both sides of Equation (15) are taken logarithmically.

$$lgS_V = (3-D)lgT_2 + (D-3)lgT_{2,max} \tag{16}$$

Therefore, the porosity fractal dimension can be obtained by taking the logarithm of the NMR technique porosity distribution curve and fitting a linear regression to Equation (16) with the slope of the regression curve as $(3-D)$.

Figure 11 represents the lgS_V and lgT_2 relationship curve, and it can be found that the curve is not a straight line, indicating that there are obvious fractal differences between micropores and macropores. A linear fit was performed for the two parts with different numbers of freeze–thaw cycles, and it was found that the linear regression fit was better. D_{min} and D_{max} were used to represent the fractal dimensions of micropores and macropores, respectively, and the results are shown in Table 3. As can be seen in Figure 12, with the increase in the number of freeze–thaw cycles, D_{max} almost did not change excessively, while D_{min} produced a certain float. This indicates that the expansion under the action of a freeze–thaw force has a greater effect on micropores compared with macropores, so the study of the evolution of the freeze–thaw deterioration of sandstone needs to consider the state of micropore distribution and micropore complexity.

Table 3. NMR fractal dimension.

Fractal Dimension	Number of Freeze–Thaw Cycles				
	0	20	40	60	80
D_{min}	0.629	0.547	0.629	0.547	0.628
D_{max}	2.981	2.978	2.981	2.978	2.981

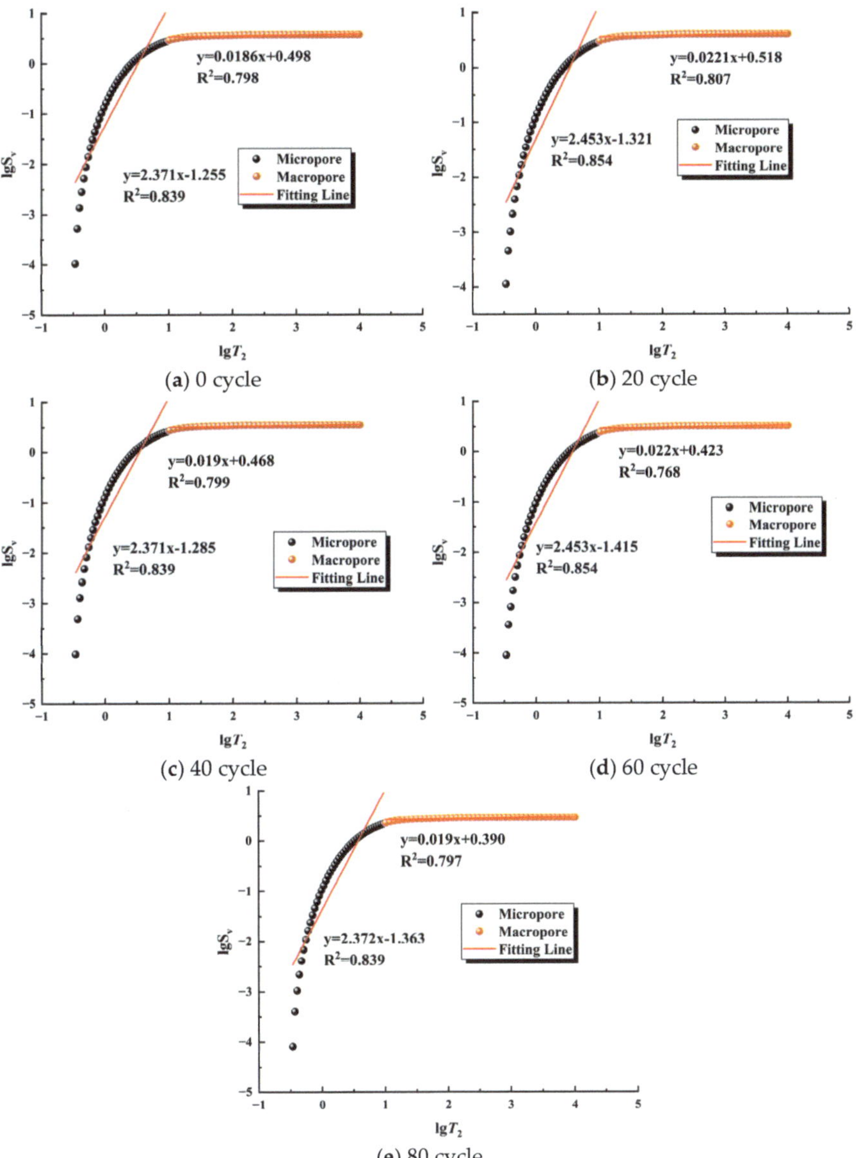

Figure 11. NMR fractal characteristics of the samples.

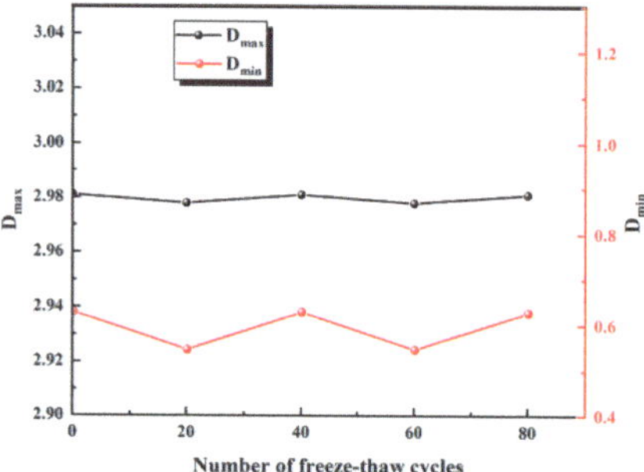

Figure 12. Changes in D_{max} and D_{min} under the freeze–thaw cycle.

Because the size of sandstone pores, compared with other hard rock pore sizes, is larger, the most obvious effect of freeze–thaw on sandstone is the change in porosity. The freeze–thaw factor in Equation (17) can be used to describe the sandstone pore destruction process.

$$W_t = \frac{1 - P_N}{1 - P_0} \tag{17}$$

where W_t is the freeze–thaw damage factor. Only the freeze–thaw damage factor can not describe the sandstone freeze–thaw damage porosity change characteristics (here, the introduction of coefficient γ), so the freeze–thaw damage factor W_t can be expressed as

$$W = \gamma W_t \tag{18}$$

$$\gamma = 1 - \frac{D_{0,min}}{D_{N,min}} \tag{19}$$

In order to investigate the relationship between the peak intensity correlation coefficient and the freeze–thaw damage factor, the model studied by Gao, F et al. [36] was fitted.

$$\frac{\sigma_N}{\sigma_0} = \eta exp(-\rho \Delta W) \tag{20}$$

where η and ρ are the correlation coefficients and ΔW is the change in the freeze–thaw damage factor $\Delta W = W_N - W_0$.

As can be seen from the Figure 13, the two show a pattern of exponential function growth, and R^2 equals 0.970. This is in high agreement with the model [36] and η equals 39.15 and ρ equals $-1/24.17$. Therefore, the use of porosity to evaluate freeze–thaw cycling damage in sandstone is feasible and applicable to other rocks.

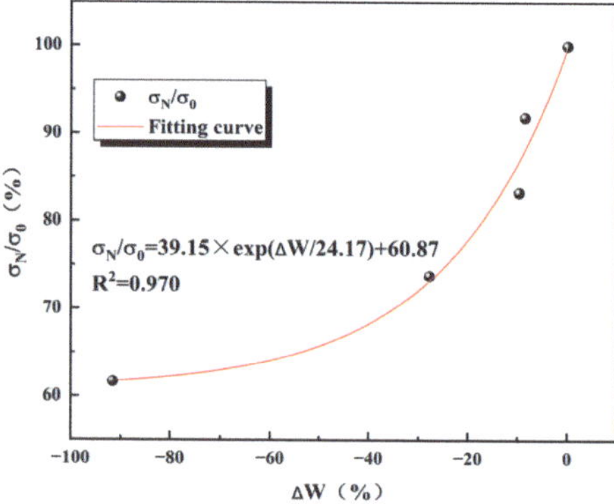

Figure 13. Peak intensity correlation coefficients versus the amount of freeze–thaw variation.

3.3.2. Scanning Electron Microscope

Sandstones are composed of quartz, sodium feldspar, calcite, hematite and other minor components, and they generally have a porosity of about 10~25% [37]. Fragments of sandstone from the unfrozen–thawed, 40 cycles of freeze–thawing and 80 cycles of freeze–thawing were taken, and the internal microstructure of the sandstone was observed at magnifications of $50\times$, $200\times$ and $500\times$; see Figure 14. There is a more pronounced change in freeze–thaw damage. It was observed that the sandstone without freeze–thawing was dense and intact, with few holes and micro-fractures visible. At $50\times$ magnification, a few micro-pores and micro-fissures are seen in the sandstone with 40 freeze–thaw cycles; at $200\times$ magnification, obvious through-fissures and pores are observed, and at $500\times$ magnification, larger pores and large fissures are found. At $50\times$ magnification, the sandstone with 80 freeze–thaw cycles has obvious pores and through-fissures compared to the sandstone with no freeze–thaw and 40 freeze–thaw cycles; at $200\times$ magnification, as the number of freeze–thaw cycles increases, more localized pores and micro-fissures appear, with a loose structure forming a honeycomb; at $500\times$ magnification, obvious through-fissures and a few micro-fissures are observed, with a large number of micro-fissures gradually developing to form a large number of micro-fractures that gradually develop into large fissures and extend, expand and penetrate to the periphery. The reason for this is that the water–ice phase change and repeated freezing and melting have weakened the cementation between the sandstone grains and damaged the internal microstructure and integrity of the sandstone, and the damage is mainly in the form of cross-grain, along-grain and tangential fractures.

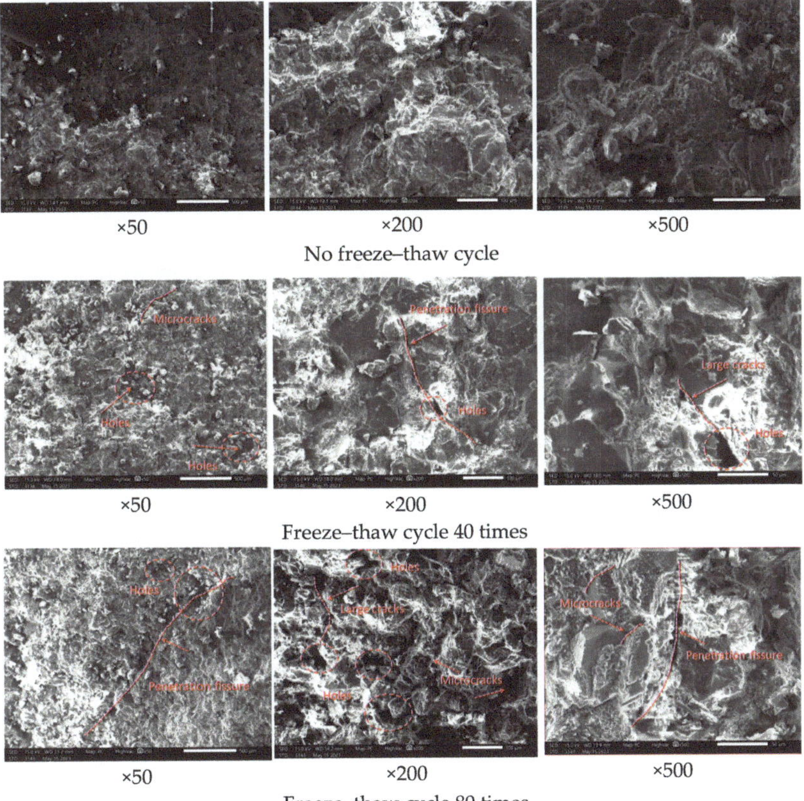

Figure 14. SEM under different numbers of freeze–thaw cycles.

3.4. Damage Evolution

3.4.1. Damage Model under the Fatigue Loading of Rock

The Weibull probability density function is introduced by statistically modeling the damage based on various defects within the rock.

$$P(\varepsilon) = \frac{m}{F}\left(\frac{\varepsilon}{F}\right)^{m-1} exp\left[-\left(\frac{\varepsilon}{F}\right)^m\right] \tag{21}$$

In the form, $P(\varepsilon)$ and ε are the internal probability distribution and microstrain of the rock under micro-stress, respectively, and m and F are the distribution parameters.

The micro-cracks within the rock under external loading N_s occur in large numbers.

$$N_s = \int_0^\varepsilon NP(x)dx = N\left\{1 - exp\left[-\left(\frac{\varepsilon}{F}\right)^m\right]\right\} \tag{22}$$

Damage to the rock is the cause of an increasing number of micro-cracks in the interior, as a consequence of defining a damage model that varies between 0 and 1.

$$D = \frac{N_s}{N} \tag{23}$$

The damage evolution equation is obtained by substituting Equation (22) into Equation (23).

$$D = 1 - exp\left[-\left(\frac{\varepsilon}{F}\right)^m\right] \tag{24}$$

The theory of loading and unloading response ratios proposed by Yin et al. [38] can be used to study the precursors of instability in nonlinear systems with the following equation.

$$Y_E = \frac{E^+}{E^-} \tag{25}$$

where E and F are the modulus of elasticity in the loading phase and the modulus of elasticity in the unloading phase, respectively.

According to Shi et al. [39], Equation (25) can be further optimized into the following equation.

$$Y_E = \frac{E^+}{E^-} = \frac{1}{1 - \frac{\varepsilon}{(1-D)}\frac{dD}{d\varepsilon}} \tag{26}$$

The derivation of Equation (24) yields

$$\frac{dD}{d\varepsilon} = (D-1)\left[-\frac{m}{F}\left(\frac{\varepsilon}{F}\right)^{m-1}\right] \tag{27}$$

The damage within the rock accumulates as it is loaded and unloaded, so the damage variables are also cumulative, and substituting Equation (27) into Equation (26) yields

$$D = \sum_{i=1}^n D_i = \sum_{i=1}^n \left(1 - e^{\frac{Y_{E(i)}-1}{mY_{E(i)}}}\right) \tag{28}$$

Based on the basic principles of damage mechanics, the rock damage model equations are as follows:

$$\sigma = E\varepsilon(1-D) \tag{29}$$

Substituting Equation (21) into Equation (25) and deriving it yields

$$\frac{d\sigma}{d\varepsilon} = Ee^{-(\frac{\varepsilon}{F})m}\left[1 - \frac{\varepsilon m}{F}\left(\frac{\varepsilon}{F}\right)^{m-1}\right] \tag{30}$$

when $\sigma = \sigma_p$, $\varepsilon = \varepsilon_p$; therefore, $\frac{d\sigma}{d\varepsilon} = 0$.

σ_p and ε_p are the peak strength and peak strain, respectively. Accordingly, the parameter F is obtained.

$$F = \frac{\varepsilon_p}{\frac{1}{m}^{\frac{1}{m}}} \tag{31}$$

Substituting $\sigma = \sigma_p$ and $\varepsilon = \varepsilon_p$ into Equation (29) yields

$$\sigma_p = E\varepsilon_p(1 - D) = E\varepsilon_p e^{-\frac{1}{m}} \tag{32}$$

Combining Equations (31) and (32) yields the following equation:

$$m = -\frac{1}{\ln\frac{\sigma_p}{E\varepsilon_p}} \tag{33}$$

$$F = \frac{\varepsilon_p}{\left(-\ln\frac{\sigma_p}{E\varepsilon_p}\right)^{-\ln\frac{\sigma_p}{E\varepsilon_p}}} \tag{34}$$

3.4.2. Damage Model under the Coupled Freeze–Thaw of Rock

The internal structure of the rock is damaged by the freezing and swelling forces, weakening the rock's ability to withstand them. Because of the complexity of the microscopic mechanisms of freeze–thaw damage, the extent of freeze–thaw damage is currently only reflected from a macroscopic perspective. The modulus of elasticity of rock is a good indicator of the deformation of rock and its ability to resist external forces, so it is usually used as a criterion for judgement.

$$D_n = 1 - \frac{E_n}{E_0} \tag{35}$$

where D_n is the freeze–thaw damage factor and E_0 and E_n are the moduli of elasticity of the rock before and after freeze–thawing, respectively.

3.4.3. Damage Model under the Coupled Freeze–Thaw-Fatigue Loading of Rocks

According to the Lemaitre strain equivalence principle, freeze–thaw damage to rock is regarded as the first damage state, and freeze–thaw damage and fatigue load damage to rock are regarded as the second damage state. Then, the two damage constitutive equations are as follows:

$$\sigma_n = (1 - D_n)E\varepsilon_n \tag{36}$$

$$\sigma = (1 - D)E_n\varepsilon \tag{37}$$

The principal structure relationship under coupled freeze–thaw-fatigue loading is obtained by combining Equation (36) and Equation (37).

$$\sigma = (1 - D)(1 - D_n)E_0\varepsilon \tag{38}$$

The damage model under coupled freeze–thaw-fatigue loading is obtained from Equation (31).

$$D_t = D + D_n - DD_n \tag{39}$$

where D_t is the damage factor under coupled freeze-thaw-fatigue loading; D is the damage factor under fatigue loading; D_n is the freeze-thaw damage factor.

4. Conclusions

The mechanical properties of sandstone under coupled freeze–thaw-fatigue action were analyzed and a damage model was developed. The studies of the microstructural changes in freeze–thaw damage led to the following main conclusions:

(1) The higher the number of freeze–thaw cycles, the lower the peak strength, frost resistance, modulus of elasticity, modulus of deformation and damping ratio; as the load cycle level increases, the modulus of deformation and modulus of elasticity increase non-linearly, the rate of increase gradually decreases, the dissipation energy due to hysteresis gradually increases and the rate of increase becomes faster and faster, while the overall damping ratio shows a pattern of gradually decreasing and increasing at a later stage.

(2) As the number of freeze–thaw cycles increases, the total porosity and micro-porosity of the sandstone increase almost linearly, and the micro-porosity is more sensitive to the effects of freeze–thaw, shifting to medium and large pores, and it is found by SEM that the higher the number of freeze–thaw cycles of the sandstone, the more micro-fractures and pores develop and the more loose the structure is, and the whole is in the shape of a nesting bee.

(3) Based on the sandstone pore fractal theory, it is found that D_{min} is more sensitive to freeze–thaw; thus, the study of freeze–thaw damage evolution law needs to consider the micro-pore distribution characteristics as well as the complexity, and based on the loading and unloading response ratio theory and strain equivalence principle, a damage model under coupled freeze–thaw-fatigue loading was established.

Author Contributions: Conceptualization, T.L. and J.H.; formal analysis, W.C. and T.L.; writing—original draft, J.H. and Y.S.; writing—review and editing, T.L. and W.C. All authors have read and agreed to the published version of the manuscript.

Funding: This work was financially supported by the National Natural Science Foundation of China under Grant No. 52374108.

Institutional Review Board Statement: Not applicable.

Informed Consent Statement: Not applicable.

Data Availability Statement: Data are contained within the article.

Conflicts of Interest: Authors Taoying Liu, Jun Huang were employed by the company Guangxi Liubin Expressway Construction and Development Co., Ltd. The remaining authors declare that the research was conducted in the absence of any commercial or financial relationships that could be construed as a potential conflict of interest.

References

1. Liu, H.; Yuan, X.; Xie, T. A damage model for frost heaving pressure in circular rock tunnel under freezing-thawing cycles. *Tunn. Undergr. Space Technol. Inc. Trenchless Technol. Res.* **2019**, *83*, 401–408. [CrossRef]

2. Deprez, M.; De Kock, T.; De Schutter, G.; Cnudde, V. A review on freeze-thaw action and weathering of rocks. *Earth-Sci. Rev.* **2020**, *203*, 103143. [CrossRef]

3. Javad, Y.; Hongyuan, L.; Andrew, C.; Fukuda, D. Experimental and numerical studies on failure behaviours of sandstones subject to freeze-thaw cycles. *Transp. Geotech.* **2021**, *31*, 100655.

4. Liu, T.; Zhang, C.; Cao, P.; Zhou, K. Freeze-thaw damage evolution of fractured rock mass suing nuclear magnetic resonance technology. *Cold Reg. Sci. Technol.* **2020**, *170*, 102951. [CrossRef]

5. Chen, Y.; Lin, H. Deterioration laws of Hoek-Brown parameters in freeze-thaw multi-fractured rock mass. *Theor. Appl. Fract. Mech.* **2022**, *123*, 103716. [CrossRef]

6. Liu, Y.; Cai, Y.; Huang, S.; Guo, Y.; Liu, G. Effect of water saturation on uniaxial compressive strength and damage degree of clay-bearing sandstone under freeze-thaw. *Bull. Eng. Geol. Environ.* **2020**, *79*, 2021–2036. [CrossRef]

7. Niu, C.; Zhu, Z.; Zhou, L.; Li, X.; Ying, P.; Dong, Y.; Deng, S. Study on the microscopic damage evolution and dynamic fracture properties of sandstone under freeze-thaw cycles. *Cold Reg. Technol.* **2021**, *191*, 103328. [CrossRef]

8. Li, J.; Zhou, K.; Ke, B. Association analysis of pore development characteristics and uniaxial compressive strength property of granite freezing-thawing cycles. *J. China Coal Soc.* **2015**, *40*, 1783–1789.

9. Li, J.; Zhou, K.; Liu, W.; Deng, H. NMR research on deterioration characteristics of microscopic structure of sandstones in freeze-thaw cycles. *Trans. Nonferrous Met. Soc. China* **2016**, *26*, 2997–3003. [CrossRef]
10. Li, J.L.; Zhu, L.Y.; Zhou, K.P.; Liu, H.W.; Cao, S.P. Damage characteristics of sandstone pore structure under freeze-thaw cycles. *Rock Soil Mech.* **2019**, *40*, 3524–3532.
11. Blykers, B.K.; Organista, C.; Kagias, M.; Marone, F.; Stampanoni, M.; Boone, M.N.; Cnudde, V.; Aelterman, J. Exploration of the X-ray Dark-Field Signal in Mineral Building Materials. *J. Imaging* **2022**, *8*, 282. [CrossRef] [PubMed]
12. Abdolghanizadeh, K.; Hosseini, M.; Saghafiyazdi, M. Effect of freezing temperature and number of freeze-thaw cycles on mode I and mode II fracture toughness of sandstone. *Theor. Appl. Fract. Mech.* **2020**, *105*, 102428. [CrossRef]
13. Wang, Y.; Yi, X.; Li, P.; Cai, M.; Sun, T. Macro-meso damage cracking and volumetric dilatancy of fault block-in-matrix induced by freeze-thaw-multistage constant amplitude cyclic (F-T-MSCAC) loads. *Fatigue Fract. Eng. Mater. Struct.* **2022**, *45*, 2990–3008. [CrossRef]
14. Liu, H.; Xu, Y.; Yang, G.; Pan, P.; Jin, L.; Tang, L.; Huang, H. Numerical experimental study on failure process of sandstone containing nature damage under freeze-thaw cycle. *J. Glaciol. Geocryol.* **2022**, *44*, 1875–1886.
15. Liu, X.; Chai, Z.; Liu, X.; Yang, Z.; Xin, Z. Pore fracture propagation and unloading failure characteristics of siltite under cyclic loads. *J. China Coal Soc.* **2022**, *47*, 77–89.
16. Wang, T.; Wang, C.; Xue, F.; Wang, L.; Xue, M. Study on acoustic and strain field evolution of red sandstone under different loading and unloading paths. *Chin. J. Rock Mech. Eng.* **2022**, *41*, 2881–2891.
17. Peng, K.; Wang, Y.; Zou, Q.; Liu, Z.; Mou, J. Effect of crack angles an energy characteristics of sandstones under a complex stress path. *Eng. Fract. Mech.* **2019**, *218*, 106577. [CrossRef]
18. Wang, J.; Li, J.; Shi, Z.; Chen, J. Energy evolution and failure characteristics of red sandstone under discontinuous multilevel fatigue loading. *Int. J. Fatigue* **2022**, *160*, 106830. [CrossRef]
19. Wang, J.; Li, J.; Shi, Z.; Chen, J.; Lin, H. Fatigue characteristics and fracture behaviour of sandstone under discontinuous multilevel constant-amplitude fatigue disturbance. *Eng. Fract. Mech.* **2022**, *274*, 108773. [CrossRef]
20. Shi, Z.; Li, J.; Wang, J. Effect of creep load on fatigue behavior and acoustic emission characteristics of sandstone containing pre-existing crack during. *Theor. Appl. Fract. Mech.* **2022**, *119*, 103296. [CrossRef]
21. Shi, Z.; Li, J.; Wang, J.; Chen, J. Fracture behavior and damage evolution model of sandstone containing a single pre-existing flaw under discontinuous fatigue loading. *Int. J. Damage Mech.* **2023**, *32*, 3–27. [CrossRef]
22. *SL/T 264-2020*; Code for Rock Tests in Water and Hydropower Projects. Ministry of Water Resources of the People's Republic of China: Beijing, China, 2020.
23. Wang, H. Mechanical Properties and Fatigue Constitutive Model of Argillaceous Quartz Siltstone under Cyclic Loading. Ph.D. Thesis, University of Science and Technology Beijing, Beijing, China, 2023.
24. Li, X.; Yao, Z.; Huang, X.; Liu, Z.; Zhao, X.; Mu, K. Investigation of deformation and failure characteristics and energy evolution of sandstone cyclic loading and unloading. *Rock Soil Mech.* **2021**, *42*, 1693–1704.
25. Yahaghi, J.; Liu, H.; Chan, A.; Fukuda, D. Experimental, theoretical and numerical modelling of the deterioration and failure process of sandstones subject to freeze–Thaw cycles. *Eng. Fail. Anal.* **2022**, *141*, 106686. [CrossRef]
26. Dongjie, H.; Qinghui, J. A general equivalent continuum model and elastic wave velocity analysis of jointed rock masses. *Int. J. Rock Mech. Min. Sci.* **2023**, *170*, 105500.
27. Zeng, X.; Chen, H.; Hu, Y. *Engineering Elastoplastic Mechanics*; Sichuan University Press: Chengdu, China, 2013; Volume 7, p. 254.
28. Zou, Q.; Chen, Z.; Zhan, J.; Chen, C.; Gao, S.; Kong, F.; Xia, X. Morphological evolution and flow conduction characteristics of fracture channels in fractured sandstone under cyclic loading and unloading. *Int. J. Min. Sci. Technol.* **2023**, *33*, 1527–1540. [CrossRef]
29. He, M.; Li, N.; Chen, Y.; Zhu, C. Damping ratio and damping coefficient of rock under different cyclic loading conditions. *Rock Soil Mech.* **2017**, *38*, 2531–2538.
30. Liu, H.; Bie, P.; Li, X.; Wei, Y.; Wang, M. Mechanical properties and energy dissipation characteristics of phyllite under triaxial multi-stage cyclic loading and unloading conditions. *Rock Soil Mech.* **2022**, *43*, 265–274+281.
31. Yang, B.; He, M.; Chen, Y. Experimental study of nonlinear damping characteristics on granite and red sandstone under the multi-level cyclic loading-unloading triaxial compression. *Arab. J. Geosci.* **2020**, *13*, 211–213. [CrossRef]
32. He, Y.; Mao, Z.; Xiao, L. A new method for constructing capillary pressure curves by using T2 distribution of NMR. *J. Univ.* **2005**, *35*, 177–181.
33. Yan, J.; Wen, D.; Li, Z.; Bin, G.; Jin-Gong, C.; Qiang, L.; Yu, Y. Quantitative evaluation method for pore structure of low permeability sandstone based on NMR logging—Taking sand four segments of south slope of Dongying sag as an example. *J. Geophys.* **2016**, *59*, 1543–1552.
34. Zhang, C.; Liu, T.; Jiang, C.; Chen, Z.; Zhou, K.; Chen, L. The Freeze-Thaw Strength Evolution of Fiber-Reinforced Cement Mortar Based on NMR and Fractal Theory: Considering Porosity and Pore Distribution. *Materials* **2022**, *15*, 7316. [CrossRef] [PubMed]
35. Porteneuve, C.; Korb, J.P.; Petit, D.; Zanni, H. Structure–texture correlation in ultra-high-performance concrete: A nuclear magnetic resonance study. *Cem. Concr. Res.* **2002**, *32*, 97–101. [CrossRef]
36. Gao, F.; Xiong, X.; Zhou, K.P.; Li, J.L.; Shi, W.C. Strength deterioration model of water-saturated sandstone under freeze-thaw cycle. *Rock Soil Mech.* **2019**, *40*, 926–932.

37. Matheus, F.E.; Mariana, G.D.R.; Rodrigo, F.B.N.; da Silva, A.F.; da Silva, M.N.P.; Tirapu-Azpiroz, J.; Lucas-Oliveira, E.; de Araújo Ferreira, A.G.; Soares, R.; Eckardt, C.B.; et al. Full scale, microscopically resolved tomographies of sandstone and carbonate rocks augmented by experimental porosity and permeability values. *Sci. Data* **2023**, *10*, 368.
38. Yin, X.; Wang, Y.; Peng, K.; Bai, Y. Development of a new approach to earthquake prediction: Load/unload response ratio (LURR) theory. *Pure Appl. Geophys.* **2000**, *157*, 2365–2383. [CrossRef]
39. Shi, Z.; Li, J.; Wang, J. Research on the fracture mode and damage evolution model of sandstone containing pre-existing crack under different stress paths. *Eng. Fract. Mech.* **2022**, *264*, 108299. [CrossRef]

MDPI

Article

Degradation Behavior and Lifetime Prediction of Polyurea Anti-Seepage Coating for Concrete Lining in Water Conveyance Tunnels

Chengcheng Peng [1], Jie Ren [2,*] and Yuan Wang [2]

1 College of Civil and Transportation Engineering, Hohai University, Nanjing 210024, China
2 College of Water Conservancy and Hydropower Engineering, Hohai University, Nanjing 210024, China
* Correspondence: rj@hhu.edu.cn

Abstract: In the lining of water conveyance tunnels, the expansion joint is susceptible to leakage issues, significantly impacting the long-term safety of tunnel operations. Polyurea is a type of protective coating commonly used on concrete surfaces, offering multiple advantages such as resistance to seepage, erosion, and wear. Polyurea coatings are applied by spraying them onto the surfaces of concrete linings in water conveyance tunnels to seal the expansion joint. These coatings endure prolonged exposure to environmental elements such as water flow erosion, internal and external water pressure, and temperature variations. However, the mechanism of polyurea coating's long-term leakage prevention failure in tunnel operations remains unclear. This study is a field investigation to assess the anti-seepage performance of polyurea coating in a water conveyance tunnel project located in Henan Province, China. The testing apparatus can replicate the anti-seepage conditions experienced in water conveyance tunnels. An indoor accelerated aging test plan was formulated to investigate the degradation regular pattern of the cohesive strength between polyurea coating and concrete substrates. This study specifically examines the combined impacts of temperature, water flow, and water pressure on the performance of cohesive strength. The cohesive strength serves as the metric for predicting the service lifetime based on laboratory aging test data. This analysis aims to evaluate the polyurea coating's cohesive strength on the tunnel lining surface after five years of operation.

Keywords: polyurea coatings; anti-seepage; accelerated aging test; cohesive strength

Citation: Peng, C.; Ren, J.; Wang, Y. Degradation Behavior and Lifetime Prediction of Polyurea Anti-Seepage Coating for Concrete Lining in Water Conveyance Tunnels. *Materials* **2024**, *17*, 1782. https://doi.org/10.3390/ma17081782

Academic Editor: Maciej Sitarz

Received: 29 March 2024
Revised: 9 April 2024
Accepted: 10 April 2024
Published: 12 April 2024

1. Introduction

Reinforced concrete has been widely used in the construction of civil infrastructure facilities. Concrete material is prone to cracking, and the water permeability of its material affects the long-term safety of the construction [1,2]. The development and utilization of sustainable composite construction materials play a crucial role in creating buildings that are not only environmentally friendly but also offer superior energy efficiency, leading to lower operational costs [3,4]. Currently, surface protective coatings are used to improve the impermeability of concrete. Research indicates that the effectiveness of these coatings is closely related to the elasticity, anti-seepage of the coating itself, and the appropriate bond strength between the substrates [5]. There are many kinds of surface coatings for concrete, including epoxy resin, polyurethane, polyurea, and acrylic acid. Even when using the same type of material, variations in formulations by different manufacturers can lead to differences in the final protective efficacy [6,7].

Polyurea elastomer is a high molecular polymer containing urea bonds (-NHCONH-) formed by the reaction of isocyanate (A component) and amine compounds (R component). Polyurea is a block copolymer composed of alternating soft and hard segments [8]. Polyurea's excellent mechanical properties can be attributed to the presence of physical crosslinking. This crosslinking arises from intermolecular and intramolecular bidentate hydrogen bonds formed between the urea linkages [9]. Polyurea is a versatile material with a wide range of applications in construction and other industries [10].

As a new type of polymer material, polyurea exhibits excellent wear and corrosion resistance [11]. It also demonstrates strong bonding capabilities with various substrates such as steel, wood, and concrete, making it an outstanding choice for surface protective coatings [12–14]. After completing the pouring and maintenance of a concrete construction, applying a spray coating to its surface can enhance the durability of the structure. Therefore, polyurea can serve as an effective surface sealing material for expansion joints in concrete linings within water conveyance tunnels. It offers advantages such as ease of construction, anti-seepage, and adaptability to expansion joint deformations. However, practical engineering applications may encounter challenges, leading to polyurea coating failure. Research indicates that coating deterioration is often linked to environmental factors such as ultraviolet exposure, temperature fluctuations, air and water infiltration, and so on. Numerous scholars have studied the degradation mechanisms of concrete surface coatings in different environments [15–20]. The hydrothermal aging of polyurea materials is an irreversible chemical reaction between infiltrating water molecules and functional groups, leading to the cleavage of chemical bonds in polyurea elastomers [21]. High-energy ultraviolet radiation can cause the cleavage of active bonds in polymers, leading to an increase in the number of polar functional groups [19]. Although several studies have shown that the thickness of polyurea coatings can vary depending on the ambient temperature [22], polyurea still exhibits excellent performance as a waterproofing coating. Polyurea coatings with thicknesses of 2 mm and 4 mm could meet the impermeability requirements of 2 mm and 5 mm cracks under the action of 300 m head water pressure [23]. The wear process of the polyurea elastomer protective material is stable, and the wear loss is linear with the time of abrasion [24]. In summary, polyurea materials exhibit excellent mechanical properties, making them well-suited for the long-term operational environment of water conveyance tunnels.

The adhesion of the coating is very important for the coating performance [25]. Horgnies et al. [26] employed a specialized peeling method and utilized scanning electron microscopy (SEM) and Fourier transform infrared spectroscopy (FTIR) to analyze the fracture energy between high-performance concrete and polyurea coating. They found that curing time, curing method, porosity of the concrete material, and surface roughness significantly influence the bond performance. Specifically, higher concrete surface roughness and porosity lead to improved bond performance. Garbacz et al. [27] used four distinct methods to characterize the roughness of concrete surfaces and examined the correlation between surface roughness and concrete bond strength. Their findings indicate that the surface roughness of concrete treated with steel shot blasting is higher than that treated with silicon sandblasting. Additionally, they noted that apart from surface roughness, the presence of cracks and loose concrete blocks are also critical factors influencing the bonding performance. Delucchi et al. [28] conducted tests on the crack-bridging ability and anti-seepage properties of four types of concrete coatings, including epoxy and polyurethane. They also proposed two experimental methods to assess coating permeability and one for evaluating coating bridging ability. The study concludes that appropriate coatings should be selected based on the specific environmental conditions of the concrete. Significant variations in the bond strength were observed before and after the water immersion test [29]. Additionally, Nguyen et al. [30] discovered that water can lead to the debonding of organic coatings from the metal substrate. They also developed a new technique for in situ measurement of water at the interface between the organic coating and substrate. In conclusion, several factors influence the bonding properties between polyurea and concrete, such as the concrete's inherent strength, surface roughness, and the conditions at the bonding interface. Nevertheless, there remains a lack of research on the adhesion aging rule of polyurea material in the water conveyance tunnel.

In water conveyance tunnels, polyurea materials, as an anti-seepage coating on the lining surface, play an important role in protecting key parts such as structural joints.

Interface damage and debonding between the coating and the substrate are the most common failure modes of coating protection. The key factor affecting the application of polyurea coating in water conveyance tunnels is to ensure good adhesion between the polyurea coating and the concrete lining. The pull-off bond test is one of the most common portable tensile test methods for measuring bond strengths between a coating and a concrete substrate in site [31]. To accurately predict the service life of polyurea materials and ensure the safe operation of tunnels, it is necessary to establish accelerated aging tests that can simulate real aging mechanisms effectively.

In this study, based on the analysis mentioned above and considering the environment of water conveyance tunnels, the primary environmental factors influencing the aging of polyurea coating are identified as water flow, water pressure, and temperature. Currently, there is a scarcity of studies and analyses focusing on these aging factors. Therefore, it is imperative to investigate the aging patterns of cohesive polyurea coatings in relation to the environment of the water conveyance tunnel. This research will enable the reasonable prediction of the service lifetime of polyurea coatings, thereby supporting the application of polyurea anti-seepage coating in tunnels.

2. Materials and Methods

2.1. Field Investigation (In Situ Testing)

In the middle of November 2019, during the maintenance of the water supply in Zhengzhou, China, the tunnel, the polyurea material at the structural joint and anchor groove was investigated, and the bond strength between the polyurea material and the concrete lining of the tunnel was sampled and tested on site. The pull-off test is a dependable method that offers several benefits. It is a simple, reliable, and easy-to-use technique for evaluating the in situ strength of concrete and the bond strength between coatings and the concrete substrate in situ. Due to the smooth surface of polyurea, achieving a strong bond with the pull head for testing is challenging. This limitation hinders the use of adhesive failure tests to directly measure the bond strength between polyurea and the concrete matrix. During the test process, the sediment is first cleaned from the coating surface. The coating surface is then roughened using sandpaper. This roughening process ensures a strong bond between the coating and the pull head. After 24 h of adhesive curing of the pull head, the circumference of the pull head by the cutting device is used to penetrate the coating to the concrete matrix. Then, a cohesive strength measuring instrument (Proceq-Dy216) is used on the roughened surface to test the cohesive strength (Figure 1).

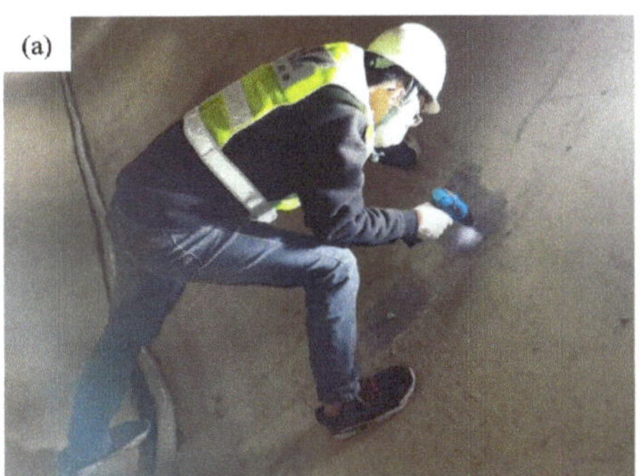

Figure 1. Pull-off test between polyurea coating and concrete lining substrates in the tunnel. (**a**) Cutting polyurea on site. (**b**) Pull-off test.

2.2. Accelerated Aging Test

2.2.1. Preparation of Specimens

The prism concrete specimens (Figure 2) were prepared with dimensions of 70 mm × 70 mm × 20 mm for the spraying polyurea coating, aiming to conduct the pull-off test. Commercial composite Portland cement P.O.42.5 was used in the specimens, with medium sand (fineness modulus between 2.3 and 3.0) and 5–10 mm diameter crushed stones mixed with pure water. The specimens underwent standard curing conditions (20 ± 2 °C, 95% relative humidity) for a duration of 28 days. The mix ratio of the concrete specimens was cement: water: sand: crushed stone: water reducer at 2.88:1:6.33:5.17:5.2%. Subsequently, the concrete surfaces were dried at 60 °C for 48 h until the moisture content dropped below 8%, following which the polyurea material was sprayed for pull-off testing (Figure 3).

Figure 2. Diagram of concrete specimen preparation.

Figure 3. Concrete specimens after spraying polyurea elastomer. (**a**) The top surface of the concrete sample sprayed with polyurea coating. (**b**) The side of the concrete sample sprayed with polyurea coating.

The experiment used polyurea materials formulated by the China Institute of Water Resources and Hydropower Research in Beijing and produced primers (SKJ-001, SKJ-002) by Qingdao Ocean New Material Technology Co., Ltd. (Qingdao, China), as well as the Spray Polyurea Elastomer (SPUA)-SKJ II polyurea coating produced by the Joint R&D Production Base of Qingdao Jialian at the Marine Chemical Research Institute. In this study, the polyurea elastomer used for spraying was synthesized as A, B dual components. Component A comprises isocyanate, while component B contains amino polyether and terminal amino chain expansion agent, tailored specifically for applications in water conveyance tunnels in Zhengzhou, Henan Province. These additives enhance the material's performance in low temperatures and humid environments and resist impact and wear. Two primers utilized were two-component silane-modified epoxy primer and polyurethane primer [32].

2.2.2. Aging Test Device

Traditional accelerated aging experiments use hydrothermal aging methods. In this study, the first step is to conduct traditional hydrothermal aging experiments using a thermostatic water bath box with a heating temperature of 20–90 °C (Figure 4a).

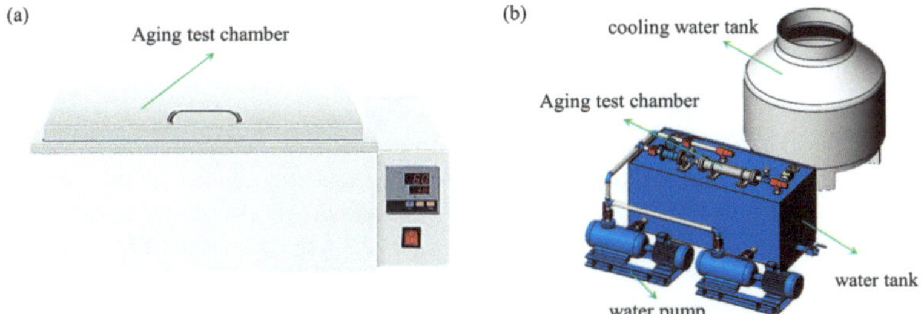

Figure 4. Schematic diagram of the test apparatus for the aging test. (**a**) Thermostatic water bath box. (**b**) Aging test equipment for simulating the operating environment of water conveyance tunnels.

The polyurea material in the water conveyance tunnel is mainly used for surface sealing and the anti-seepage effect of the joints in the lining structure. It is affected by multiple compound factors such as water flow impact, water pressure, and temperature changes in the tunnel for a long time. Therefore, this study combines the actual engineering environment and designs and manufactures an aging test device that can simulate the environmental conditions of the water conveyance tunnel site. It can be used to study the performance deterioration process of polyurea material under the combined effects of multiple factors such as temperature, water flow, and pressure in the tunnel (Figure 4b).

The apparatus replicates the conditions of a water conveyance tunnel, facilitating accelerated specimen aging through temperature elevation (Figure 4b). This device can examine the variation in cohesive strength of polyurea specimens under a composite influence of temperature, dynamic water flow, and pressure. The testing apparatus consists of a water tank, two constant water pumps, an aging test chamber, a cooling water tank, a temperature measurement, a flow meter, two pressure gauges, and a control device. To fulfill the demands of prolonged continuous operation, two constant water pumps operate alternately. A heating device and a temperature sensor were incorporated into the water tank to elevate the temperature and expedite specimen aging. Aligned with the real water flow conditions of a water conveyance tunnel in Henan Province, the test apparatus can simultaneously simulate aging environments with varying water pressures of up to 0.65 MPa (with a maximum water temperature reaching 80 °C). Concrete specimens coated with polyurea material on the surface, each measuring 70 mm by 70 mm by 20 mm, were positioned on both sides of the unit, with a polyurea material placed in the middle (Figure 5). During the experiment, water was heated to the designated temperature in a water tank. A water pump was used to continuously circulate water with a certain pressure and flow rate into the aging test chamber, simulating the operating conditions of a real water delivery tunnel.

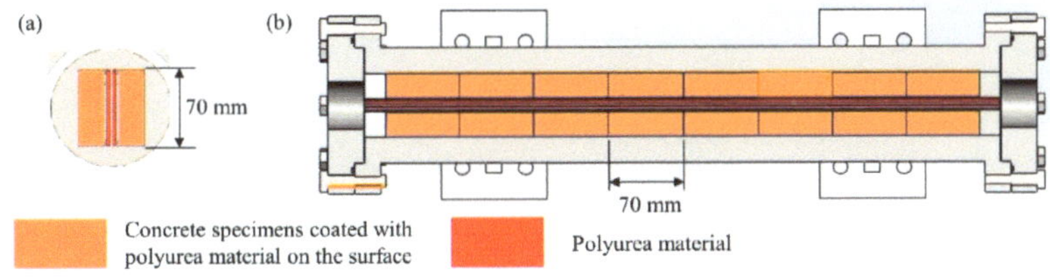

Figure 5. Schematic diagram of internal specimen placement of aging test chamber. (**a**) Left view of the aging test chamber. (**b**) Vertical view of the aging test chamber.

2.2.3. Experimental Design

In traditional accelerated aging experiments, a well-established approach to expedite the aging of polyurea coatings in the lab is to elevate the temperature. Align with the environmental conditions encountered within water conveyance tunnels, the combined effects of water and temperature aging factors were used to evaluate the cohesive strength variation. Concrete specimens coated with polyurea material were immersed in water at constant temperatures of 20 °C, 50 °C, 65 °C, and 80 °C. Subsequently, the degradation performance of cohesive strength was measured after 7-day and 21-day test cycles.

The test water temperatures were set at 50 °C, 65 °C, and 80 °C during indoor tests on the degradation behavior of polyurea materials in simulated on-site environments. The polyurea coating–concrete specimen was placed in the central slot of the primary testing apparatus to expose it to the combined effects of these factors. This configuration allows water to flow over the material's surface, simulating a hydraulic tunnel's internal water flow conditions. To prevent water erosion from undermining the bond between the polyurea and the concrete specimen, the polyurea coating was applied to the side of the specimen (Figure 3b). The aging period spans 1, 3, 7, 14, 21, 25, 27, and 28 days. Previous research indicates significant discrepancies in cohesive strength measurement results due to test instrument and method variations. Hence, to ensure comparability with field measurements, the cohesive strength of the specimen was also assessed using the same portable instrument (Prodeq-DY216) [33]. The following table (Table 1) details the various experimental designs employed for the aging test of polyurea coatings.

Table 1. Experimental design for aging test of polyurea coatings.

Experiment Type	Case	Aging Temperature (°C)	Aging Duration (d)
	Case 1-1	20	7
	Case 1-2	20	21
	Case 1-3	50	7
Hydrothermal aging test	Case 1-4	50	21
	Case 1-5	65	7
	Case 1-6	65	21
	Case 1-7	80	7
	Case 1-8	80	21
	Case 2-1	50	1
	Case 2-2	50	3
	Case 2-3	50	7
	Case 2-4	50	14
	Case 2-5	50	21
	Case 2-6	50	25
	Case 2-7	50	27
	Case 2-8	50	28
	Case 3-1	65	1
	Case 3-2	65	3
	Case 3-3	65	7
Aging test under simulated on-site environment	Case 3-4	65	14
	Case 3-5	65	21
	Case 3-6	65	25
	Case 3-7	65	27
	Case 3-8	65	28
	Case 4-1	80	1
	Case 4-2	80	3
	Case 4-3	80	7
	Case 4-4	80	14
	Case 4-5	80	21
	Case 4-6	80	25
	Case 4-7	80	27
	Case 4-8	80	28

3. Results

3.1. The Results of In Situ Testing

In Zhengzhou Province, a water conveyance tunnel had been operating successfully for five years prior to the commencement of this testing. However, recent investigations have revealed the occurrence of bulging, rupturing, and water seepage in the polyurea material. The cohesive strength of polyurea coatings was evaluated at the land and underwater sections, with three repeated tests conducted near each sampling point. The following table

(Table 2) details each sampling point. Test points 1 to 3 were chosen along the tunnel's sidewall, points 4 to 6 on the tunnel floor, and point 7 at the anchor channel. The result of in situ testing is presented in the following figure (Figure 6). The average cohesive strength recorded for the tunnel was 1.827 MPa and 1.367 MPa, with minimum values of 2.035 MPa and 1.067 MPa, respectively.

Table 2. Sampling point of in situ testing.

Case	Testing Location	Chainage of Land Sections	Chainage of Underwater Sections
1		27-28	96–97
2	The wall of the structural joint	68–69	255–256
3		174–175	342–343
4		27–28	96–97
5	The bottom of the structural joint	67–68	255–256
6		174–175	342–343
7	Anchor channel	69	342

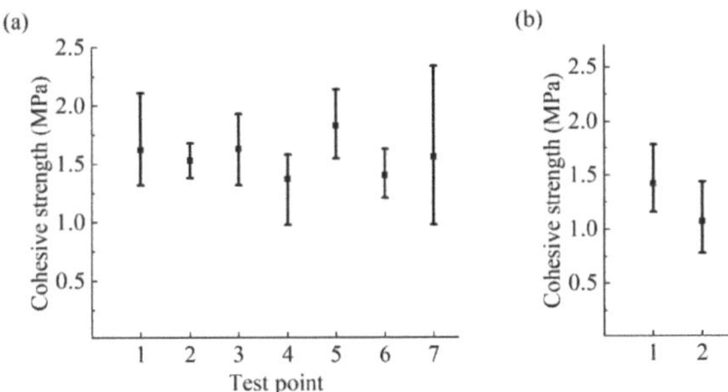

Figure 6. The cohesive strength of polyurea coating samples in Zhengzhou water conveyance tunnel. (**a**) Land sections of the water conveyance tunnel. (**b**) Underwater sections of the water conveyance tunnel.

Notably, test results at the same position exhibited significant variability. It is noted that the cohesive strength between polyurea coating and the concrete lining can reach 2.2 MPa in wet or water environments. However, a substantial decrease in cohesive strength between polyurea and the concrete lining was observed along the tunnel's length, compared to initial values, under the influence of the tunnel environment [32].

3.2. Aging Test Results in the Laboratory

3.2.1. Hydrothermal Aging Test

In this section, the results of cohesive strength between polyurea coating and concrete after the hydrothermal aging tests are presented. The initial cohesive strength values for the three sets of samples between polyurea coating and concrete substrates tested in the laboratory were 3.829 MPa, 3.330 MPa, and 3.290 MPa, with an average cohesive strength of 3.493 MPa. The specimens coated with polyurea were immersed in water baths set at different temperatures, and changes in cohesive strength were recorded (Table 3).

Table 3. Results of hydrothermal aging test.

Case	Wrapped around Polyurea				Wrapped on Both Sides of Polyurea			
	Values in MPa		Average	Standard Deviation	Values in MPa		Average	Standard Deviation
Case 1-1	*	3.72	3.72	-	3.05	3.82	3.44	0.39
Case 1-2	2.93	3.53	3.23	0.30	3.00	3.49	3.25	0.25
Case 1-3	3.28	-	3.28	-	3.84	3.38	3.61	0.23
Case 1-4	3.07	3.49	3.43	0.21	3.42	3.09	3.26	0.17
Case 1-5	2.52	2.97	2.75	0.23	3.10	2.77	2.94	0.17
Case 1-6	2.22	2.45	2.34	0.12	2.16	2.04	2.10	0.06
Case 1-7	1.63	1.50	1.57	0.06	2.18	1.92	2.05	0.13
Case 1-8	*	-	-	-	1.04	0.45	0.75	0.30

* Bonding adhesive failure at pull head.

During the testing, the impact of temperature increase was considered. This led to water vapor accumulation between the polyurea and concrete surfaces, which resulted in accelerated bonding aging, which was evaluated using two specimens maintained at the same temperature (Figure 3b). As the test water temperature increased, the decline in cohesive strength became more pronounced. Notably, except for the 80 °C water temperature condition, the cohesive strength test results differed between the two specimens subjected to the same aging conditions. Specifically, when the test water temperature reached 80 °C, the cohesive strength of the cut specimen was significantly higher than that of the uncut specimen during the same test period. This analysis suggests that the uncut specimen is more prone to water vapor accumulation at the bonding interface between polyurea and concrete under high water temperature conditions, exacerbating the aging of the bonding material.

3.2.2. Aging Test under Simulated On-Site Environment

This section presents the experimental results of aging tests under a simulated on-site environment. The effect of aging on cohesive strength is shown in the following figure (Figure 7).

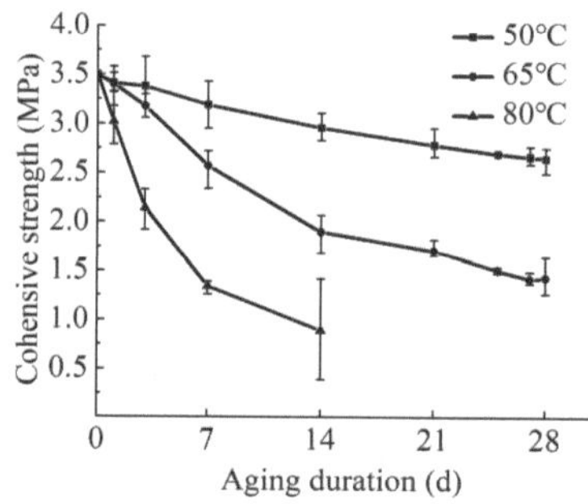

Figure 7. The variation of cohesive strength after simulated water tunnel environmental aging test over time at different temperatures.

As expected, increasing ambient temperature and test duration significantly reduced cohesive strength. This effect was particularly pronounced at 80 °C, where a rapid decline in cohesive strength was observed, potentially leading to debonding after a short test period. These findings are consistent with previous results of the hydrothermal aging test. Notably, comparative analysis revealed that cohesive strength under water flow and pressure conditions was superior to that observed in the current test, highlighting the influence of these factors.

4. Discussion

4.1. Failure Types Analysis

To aid in understanding the failure mechanisms, Figure 8 defines and summarizes the various failure types observed during the aging tests. Based on the different bond failure types observed after cohesive strength measurement, the specimens can be classified into three categories: A, B, and C. Type A damage can be summarized as cohesive failure between concrete and epoxy resin, type B damage can be summarized as interfacial failure between the epoxy resin and polyurethane primer, and type C damage can be summarized as interfacial failure between polyurea and primer.

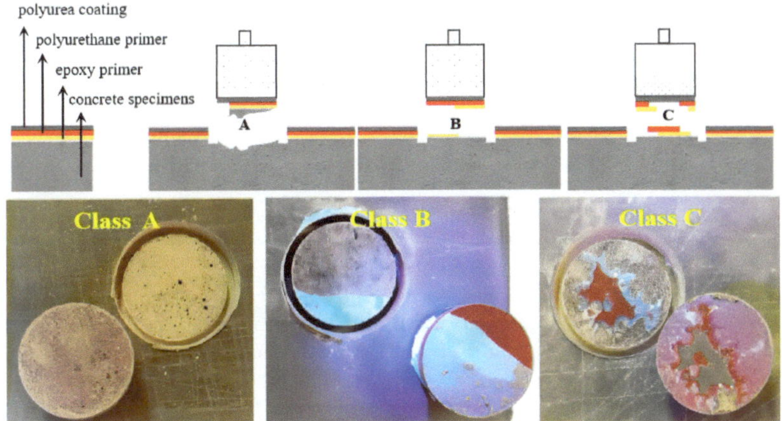

Figure 8. Specimens of the three different bond failure types.

4.1.1. Failure Types of the Hydrothermal Aging Test

This section illustrates the bond failure types of polyurea–concrete specimens after the hydrothermal aging test (Figure 9). The first row in the figure shows the specimens at a water temperature of 20 °C, the second row shows the specimens at 50 °C, the third row shows the specimens at 65 °C, and the fourth row shows the specimens at 80 °C. Under the experimental aging duration of 7 days, the failure modes of almost all specimens were of class A. Under the aging duration of 21 days, when the test water temperature was 65 and 80 °C, the failure types of the test specimens appeared to be of class B/C.

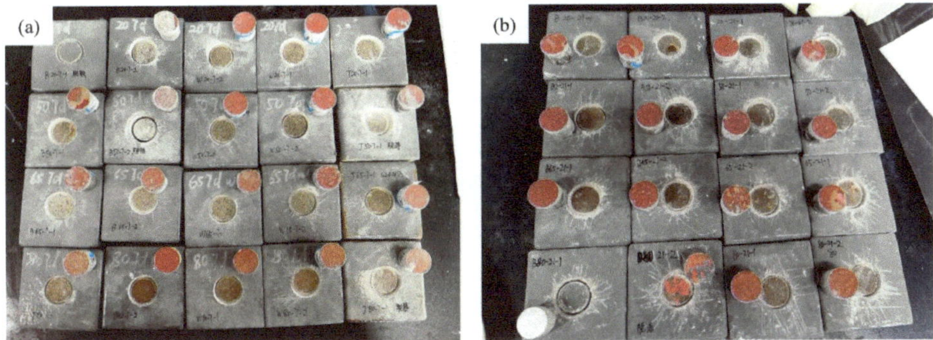

Figure 9. The failure types of polyurea–concrete specimens after the hydrothermal aging test. (**a**) Aging duration is 7 days. (**b**) Aging duration is 21 days.

Under mild aging conditions (short duration, low temperature), class A failure was observed, characterized by an adhesive failure within the concrete layer and residual concrete adhering to the epoxy surface. As aging time increased, debonding at the concrete–epoxy interface became the dominant failure mode. Notably, at 80 °C for 21 days, specimens wrapped around polyurea exhibited debonding with damage to the epoxy and polyurethane coatings. Red polyurethane primer degradation was particularly evident at 65 °C and 80 °C water temperatures, consistent with previous studies. Hydrothermal aging demonstrably weakens the adhesive performance of the epoxy layer. The observed failure progression suggests initial damage within the concrete substrates, debonding at the concrete–epoxy interface, and ultimately, with extended aging, degradation of the polyurethane primer and epoxy coating.

4.1.2. Failure Types of the Simulated Tunnel Environment Aging Test

Following the simulated tunnel environment aging test, the bond failure types of the specimens were observed to be as follows. At a test water temperature of 50 °C, the specimens exhibited the failure type of class A. At a test water temperature of 65 °C, type B failure occurred after 21 days of the aging duration. At a test water temperature of 80 °C, type B/C failure occurred after only 7 days of the aging duration.

Consistent with prior observations, at shorter aging times, failure is dominated by class A detachment, where the epoxy resin separates from the concrete base surface. As aging temperature and duration increase, the destruction mode transitions to class B and class C failures, characterized by the retention of epoxy and polyurethane undercoating on the exposed concrete surface. These findings corroborate those reported by other researchers [19,31,34]. At 80 °C water temperature, the adhesive interface degrades more rapidly due to water pressure (Figure 10). The first row in figure shows the bond failure types of specimens tested at 80 °C for 7 days. The second row shows the specimens tested for 11 and 13 days, and the third row shows the specimens tested for 14 days.We propose that water ingress along the concrete-polyurea interface triggers a reaction between the epoxy resin and polyurethane primer with water, leading to the observed bulges on some sample surfaces. By day 14, bond strength is significantly reduced, with an uneven polyurea–concrete interface exhibiting exposed gray polyurea coating in some areas.

Figure 10. The bond failure types of specimens at 80 °C water during the aging test simulated in the tunnel environment.

4.1.3. SEM Characterization of Typical Failure Types

Following pull-off strength testing, scanning electron microscopy (SEM) was employed to analyze the microscopic morphology of the fractured interface. The interfacial analysis after exposure to Case 3-8 is depicted in the following figures (Figures 11a and 12a). Post-fracture analysis revealed the presence of residual epoxy resin and polyurethane undercoating on the concrete base surface. Microscopic magnification clearly visualized the coating and fracture texture (Figure 11b). A specimen from the test Case 3-1 observed that there were obvious defects on the concrete surface (Figure 11b). Preexisting defects necessitated epoxy resin repair before spraying. The fractured interface displayed visible epoxy resin and concrete, with the repair material completely filling the defect. High magnification revealed the epoxy resin–concrete interface with small adhering concrete fragments. The bond strength measured for this repaired specimen exhibited a slight increase compared to the standard test group under the same aging conditions. These findings suggest the potential utility of epoxy resin for mitigating concrete surface defects and enhancing the contact area, thereby improving the bond strength between the epoxy resin and the concrete surface.

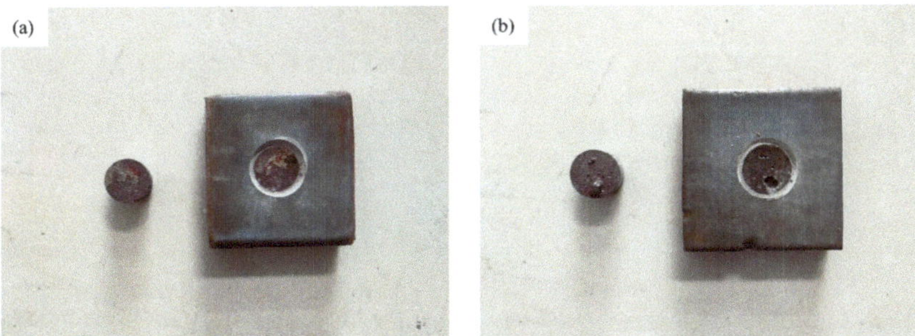

Figure 11. Macroscopic morphology of the fracture surface. (**a**) Case 3-8. (**b**) Case 3-1.

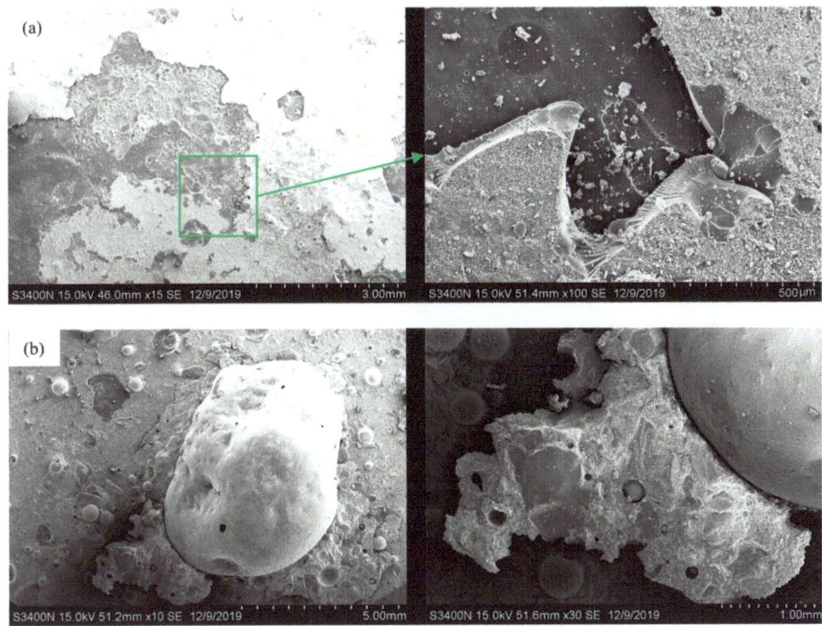

Figure 12. Micro morphology of the fracture surface. (**a**) Case 3-8. (**b**) Case 3-1.

4.2. Lifetime Prediction

This study employs artificial accelerated aging to simulate the effects of temperature, water pressure, and water flow on the cohesive strength and bond failure types between polyurea coating and concrete linings. Drawing on the test data in Section 3.2.2, this section predicts the polyurea's service life for the project to guarantee operational safety. Kinetic equations, often expressed as exponential functions (Equation (1)), have been successfully employed to describe the relationship between aging performance (P) and aging time (t) for similar polymeric materials [35]. We will utilize this approach to predict the service life of the polyurea material.

$$f(P) = e^{-Kt^{\alpha}}, \tag{1}$$

where K is the chemical reaction rate constant, unit min^{-1}, t is the time, unit day, and α is the empirical constant. In terms of cohesive strength, $f(P_\sigma) = \frac{\sigma}{\sigma_n}$, σ is cohesive strength; σ_n is the initial cohesive strength.

Differentiating Equation (1) results in Equation (2).

$$\ln(P_\sigma) = A_1 + B_1 t^\alpha \, , \tag{2}$$

Equation (2) was employed to establish a mathematical relationship between cohesive strength and aging time under varying temperature conditions within the simulated tunnel environment aging test. The following table (Table 4) presents the fitting parameters, while Figure 13 depicts the corresponding fitting curves.

Table 4. Fitting parameters at different temperatures.

Temperature/°C	Parameter			Correlation
	A1	**B1**	α	
50	0.00185 ± 0.00587	-0.01937 ± 0.00359	0.80354 ± 0.05232	0.99719
65	0.03319 ± 0.03828	-0.07949 ± 0.02575	0.74491 ± 0.09072	0.98999
80	0.03273 ± 0.08136	-0.25358 ± 0.07937	0.65825 ± 0.10981	0.98892

Given the application of polyurea coatings in water conveyance tunnels, this study acknowledges the limitations of the traditional Arrhenius lifetime model, which primarily focuses on temperature. The Eyring reaction theory model incorporates nonthermal aging factors, and its expression is as follows.

$$L(V) = \frac{1}{V} Const. \cdot e^{\frac{D}{V}} \, , \tag{3}$$

where V is the stress value in absolute units (such as relative humidity); $L(V)$ is the life scale; D is the undetermined parameters of the model.

This study integrates the Arrhenius lifetime model with the Eyring reaction theory model to account for the combined effects of water and temperature on material aging, yielding a comprehensive water–thermal aging life model [36].

$$L(H, T) = \frac{a_1}{H} e^{\frac{b_1}{H} + \frac{c_1}{T}} \, , \tag{4}$$

where $L(H,T)$ is the water–thermal aging life model, T is the thermodynamic temperature, and K, a_1, b_1, c_1 are the undetermined parameters of the model.

Following the cohesive strength results of the simulated tunnel environment aging test, when the humidity is considered, Equation (4) can be simplified to [37]:

$$L(H_0, T) = a_{H_0} e^{\frac{c_1}{T}} \, , \tag{5}$$

where $a_{H_0} = a_1 e^{\frac{b_1}{H_0}/H_0}$; H_0 is the constant relative to humidity, consistent with the Arrhenius model.

According to the specification GB/T 23446-2009 [38], the bond strength of polyurea must exceed 2.5 MPa. The critical value of bond strength was set at 50% of this value. The critical value of cohesive strength was also set at 1.25 MPa. After fitting, the parameter substitution Equation (5) was obtained, as shown in Figure 14. If the annual average temperature of the water conveyance tunnel is 15 °C, the estimated service life of the polyurea coatings is approximately 16.52 years.

$$L(H_0, T) = 4.96 \times 10^{-12} e^{\frac{10008.5612}{T}} \tag{6}$$

The water–thermal aging lifetime model suggests a service life for the polyurea coating. This is attributed to the combined effects of temperature, water flow, and other factors encountered during real-world engineering applications, as reflected by the time-dependent cohesive strength profile in Figure 15. In the water conveyance tunnel project, with a predicted temperature range of 10~30 °C and a service life of 5 years, the measured cohesive strength of the polyurea exhibited a range of 0.83~2.93 MPa. The average estimated value was 1.56 MPa, with the lowest and highest values being 1.067 MPa and 2.035 MPa, respectively. These results demonstrate good agreement with the predicted values.

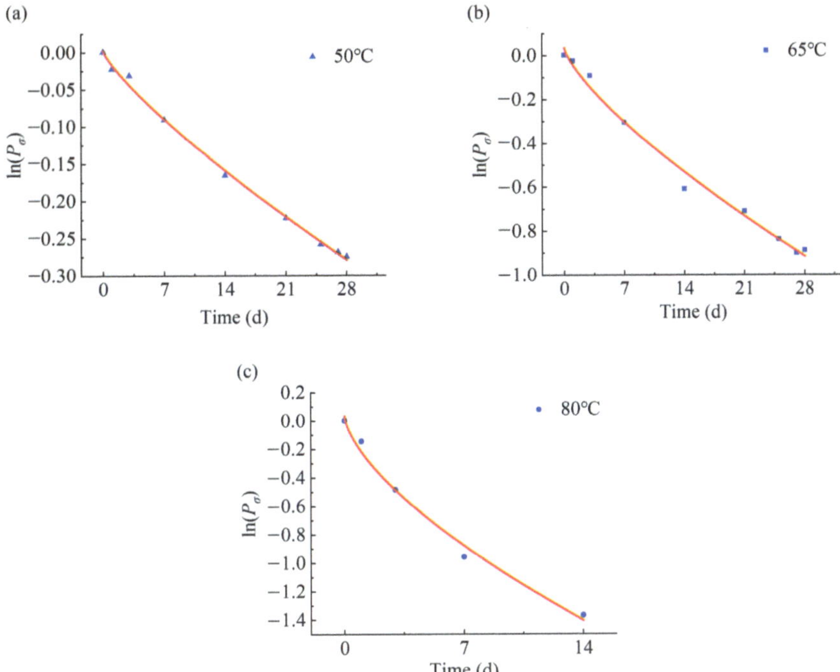

Figure 13. Relationship between cohesive strength and aging time. (**a**) Aging temperature is 50 °C. (**b**) Aging temperature is 65 °C. (**c**) Aging temperature is 80 °C.

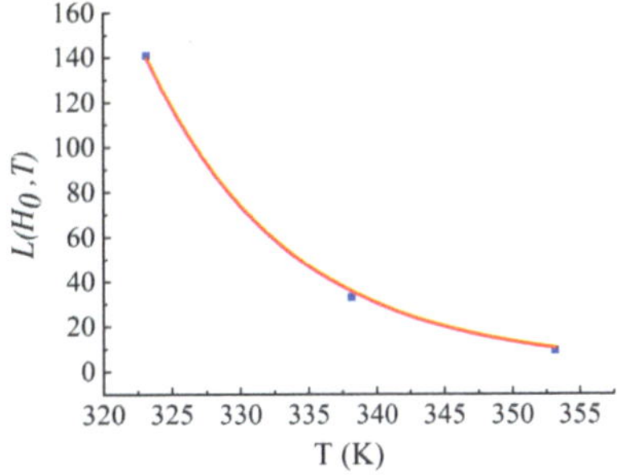

Figure 14. Fitting results of the water–thermal aging lifetime model.

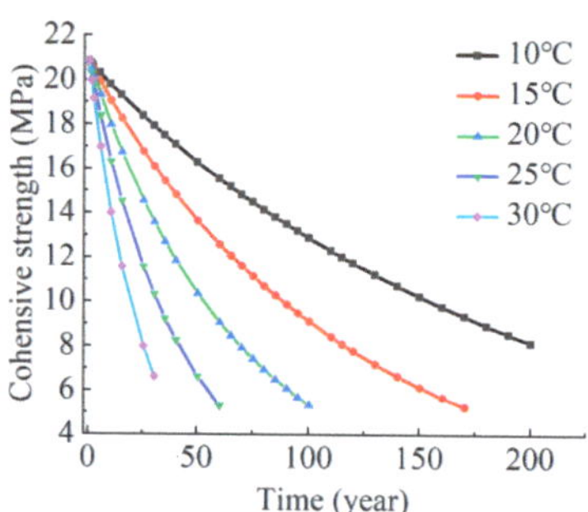

Figure 15. Predicted time-dependent cohesive strength profile.

5. Conclusions

This study investigates the degradation of polyurea-sprayed coatings used in a water conveyance tunnel in Henan, China. The results reveal significant variations in the bond strength between the polyurea coatings and the concrete substrate after five years of operation, indicating environmental influence on bond performance.

An indoor accelerated aging test was designed to replicate the actual operating conditions of the water conveyance tunnel project, including temperature, dynamic water flow, and pressure. This test aimed to elucidate the deterioration of adhesive properties between the polyurea coatings and concrete lining. Three distinct failure modes were identified based on the observed debonding locations: (A) cohesive failure between concrete and epoxy resin, (B) interfacial failure between epoxy resin and polyurethane primer, and (C) interfacial failure between polyurea and primer.

The hydrothermal aging test demonstrated a progressive decrease in cohesive strength between the polyurea and the concrete substrates with extended aging time at a constant temperature. Notably, the aging test simulating the tunnel environment revealed that class A failure dominated at shorter aging times. However, with extended aging, the failure mode transitioned to class B failure and class C failure, characterized by debonding with residual epoxy and polyurethane primer on the exposed concrete surface.

The water–thermal aging lifetime model was employed to predict the service life of the polyurea coatings in the tunnel, with a threshold cohesive strength of 1.25 MPa, signifying acceptable bond performance. Assuming an annual average water tunnel temperature of 15 °C, this model successfully captured the aging trend of the polyurea–concrete cohesive strength, demonstrating good agreement with field investigations and testing.

This study designed an aging test device that simulates the real operating environment of water delivery tunnels. Indoor accelerated experiments were conducted to study the degradation law of the bond strength between polyurea and concrete under real conditions. The reasonable service life of the polyurea–concrete waterproof system in the tunnel environment was predicted. Considering the combined effects of temperature, water flow, and water pressure on the bonding performance of polyurea, a water–thermal life prediction model was constructed to predict its service life. In the future, more aging factors such as stress in the tunnel will be considered. The aging constitutive model of polyurea will be constructed, and the prediction model of its service life will be improved to predict its service life more reasonably.

Author Contributions: Conceptualization, J.R. and Y.W.; Methodology, J.R., Y.W. and C.P.; Investigation, C.P. and J.R.; Writing, C.P. All authors have read and agreed to the published version of the manuscript.

Funding: This research was funded by the National Natural Science Foundation of China (U2240210; U23B20144; 52209129).

Institutional Review Board Statement: Not applicable.

Data Availability Statement: The data presented in this research are available on request from the corresponding author.

Conflicts of Interest: The authors declare that they have no known competing financial interests or personal relationships that could have appeared to influence the work reported in this paper.

References

1. Barbucci, A.; Delucchi, M.; Cerisola, G. Organic coatings for concrete protection: Liquid water and water vapour permeabilities. *Prog. Org. Coat.* **1997**, *30*, 293–297. [CrossRef]
2. Basheer, L.; Kropp, J.; Cleland, D.J. Assessment of the durability of concrete from its permeation properties: A review. *Constr. Build. Mater.* **2001**, *15*, 93–103. [CrossRef]
3. Kanagaraj, B.; Kiran, T.; Gunasekaran, J.; Nammalvar, A.; Arulraj, P.; Gurupatham, B.G.A.; Roy, K. Performance of Sustainable Insulated Wall Panels with Geopolymer Concrete. *Materials* **2022**, *15*, 8801. [CrossRef]
4. Roy, K.; Gurupatham, B.G.A. Editorial for the Special Issue on Sustainable Composite Construction Materials. *J. Compos. Sci.* **2023**, *7*, 491. [CrossRef]
5. Khanzadeh Moradllo, M.; Shekarchi, M.; Hoseini, M. Time-dependent performance of concrete surface coatings in tidal zone of marine environment. *Constr. Build. Mater.* **2012**, *30*, 198–205. [CrossRef]
6. Almusallam, A.F.M.K. Performance of concrete coatings under varying exposure conditions. *Mater. Struct.* **2002**, *35*, 487–494. [CrossRef]
7. Almusallam, A.A.; Khan, F.M.; Dulaijan, S.U.; Al-Amoudi, O.S.B. Effectiveness of surface coatings in improving concrete durability. *Cem. Concr. Compos.* **2003**, *25*, 473–481. [CrossRef]
8. Iqbal, N.; Tripathi, M.; Parthasarathy, S.; Kumar, D.; Roy, P.K. Polyurea spray coatings: Tailoring material properties through chemical crosslinking. *Prog. Org. Coat.* **2018**, *123*, 201–208. [CrossRef]
9. Iqbal, N.; Tripathi, M.; Parthasarathy, S.; Kumar, D.; Roy, P.K. Polyurea coatings for enhanced blast-mitigation: A review. *RSC Adv.* **2016**, *6*, 1976–19717. [CrossRef]
10. Shojaei, B.; Najafi, M.; Yazdanbakhsh, A.; Abtahi, M.; Zhang, C. A review on the applications of polyurea in the construction industry. *Polym. Advan Technol.* **2021**, *32*, 2797–2812. [CrossRef]
11. Zur, E. Polyurea—The New Generation of Lining and Coating. *Adv. Mater. Res.* **2010**, *95*, 85–86. [CrossRef]
12. Toutanji, H.A.; Choi, H.; Wong, D.; Gilbert, J.A.; Alldredge, D.J. Applying a polyurea coating to high-performance organic cementitious materials. *Constr. Build. Mater.* **2013**, *38*, 1170–1179. [CrossRef]
13. Alldredge, D.J.; Gilbert, J.A.; Toutanji, H.; Lavin, T.; Balasubramanyam, M. *Structural Enhancement of Framing Members Using Polyurea*; Springer New York: New York, NY, USA, 2011; pp. 49–63. ISBN 2191-5644.
14. Ha, S.K.; Lee, H.K.; Kang, I.S. Structural behavior and performance of water pipes rehabilitated with a fast-setting polyurea–urethane lining. *Tunn. Undergr. Sp. Tech.* **2016**, *52*, 192–201. [CrossRef]
15. Liu, F.; Yin, M.; Xiong, B.; Zheng, F.; Mao, W.; Chen, Z.; He, C.; Zhao, X.; Fang, P. Evolution of microstructure of epoxy coating during UV degradation progress studied by slow positron annihilation spectroscopy and electrochemical impedance spectroscopy. *Electrochim. Acta* **2014**, *133*, 283–293. [CrossRef]
16. Yang, X.F.; Tallman, D.E.; Bierwagen, G.P.; Croll, S.G.; Rohlik, S. Blistering and degradation of polyurethane coatings under different accelerated weathering tests. *Polym. Degrad. Stabil.* **2002**, *77*, 103–109. [CrossRef]
17. Youssef, G.; Whitten, I. Dynamic properties of ultraviolet-exposed polyurea. *Mech. Time-Depend. Mat.* **2017**, *21*, 351–363. [CrossRef]
18. Le Gac, P.Y.; Choqueuse, D.; Melot, D. Description and modeling of polyurethane hydrolysis used as thermal insulation in oil offshore conditions. *Polym. Test.* **2013**, *32*, 1588–1593. [CrossRef]
19. Wang, H.; Feng, P.; Lv, Y.; Geng, Z.; Liu, Q.; Liu, X. A comparative study on UV degradation of organic coatings for concrete: Structure, adhesion, and protection performance. *Prog. Org. Coat.* **2020**, *149*, 105892. [CrossRef]
20. Kumano, N.; Mori, K.; Kato, M.; Ishii, M. Degradation of scratch resistance of clear coatings by outdoor weathering. *Prog. Org. Coat.* **2019**, *135*, 574–581. [CrossRef]
21. Pal, G.; Al-Ostaz, A.; Li, X.; Alkhateb, H.; Cheng, A.H.D. Molecular Dynamics Simulations and Hygrothermal Aging of Polyurea–POSS Nanocomposites. *J. Polym. Environ.* **2015**, *23*, 171–182. [CrossRef]
22. Wang, Z. Study on the variation law of the thickness of spraying quick-setting waterproof coating materials with time and temperature. *IOP Conf. Ser. Earth Environ. Sci.* **2019**, *267*, 32057. [CrossRef]
23. Li, B.; Li, J.; Yin, X. Study on the Performance of Polyurea Anti-Seepage Spray Coating for Hydraulic Structures. *Sustainability* **2023**, *15*, 9863. [CrossRef]
24. Wang, X.; Luo, S.; Liu, G.; Zhang, L.C.; Wang, Y. Abrasion test of flexible protective materials on hydraulic structures. *Water Sci. Eng.* **2014**, *7*, 106–116. [CrossRef]
25. Lyon, S.B.; Bingham, R.; Mills, D.J. Advances in corrosion protection by organic coatings: What we know and what we would like to know. *Prog. Org. Coat.* **2017**, *102*, 2–7. [CrossRef]
26. Horgnies, M.; Willieme, P.; Gabet, O. Influence of the surface properties of concrete on the adhesion of coating: Characterization of the interface by peel test and FT-IR spectroscopy. *Prog. Org. Coat.* **2011**, *72*, 360–379. [CrossRef]

27. Garbacz, A.; Courard, L.; Kostana, K. Characterization of concrete surface roughness and its relation to adhesion in repair systems. *Mater. Charact.* **2006**, *56*, 281–289. [CrossRef]

28. Delucchi, M.; Barbucci, A.; Cerisola, G. Crack-bridging ability and liquid water permeability of protective coatings for concrete. *Prog. Org. Coat.* **1998**, *33*, 76–82. [CrossRef]

29. Negele, O.A.W.F. Internal Stress and Wet Adhension of Organic Coatings. *Prog. Org. Coat.* **1996**, *28*, 285–289. [CrossRef]

30. Nguyen, T.; Byrd, E.; Bentz, D.; Lint, C. In situ measurement of water at the organic coating/substrate interface. *Prog. Org. Coat.* **1996**, *27*, 181–193. [CrossRef]

31. Fazli, H.; Yassin, A.Y.M.; Shafiq, N.; Teo, W. Pull-off testing as an interfacial bond strength assessment of CFRP-concrete interface exposed to a marine environment. *Int. J. Adhes. Adhes.* **2018**, *84*, 335–342. [CrossRef]

32. Li, B.; Zhou, Y.; Xiao, J.; Li, X. Study on the new polyurea-composite impervious system of expansion joints under reverse pressure. *Shuili Xuebao* **2015**, *46*, 1479–1486. (In Chinese) [CrossRef]

33. Momayez, A.; Ehsani, M.R.; Ramezanianpour, A.A.; Rajaie, H. Comparison of methods for evaluating bond strength between concrete substrate and repair materials. *Cem. Concr. Res.* **2005**, *35*, 748–757. [CrossRef]

34. Mayer-Trzaskowska, P.; Robakowska, M.; Gierz, Ł.; Pach, J.; Mazur, E. Observation of the Effect of Aging on the Structural Changes of Polyurethane/Polyurea Coatings. *Polymers* **2024**, *16*, 23. [CrossRef] [PubMed]

35. Li, Y.J. Study on Degradation Rule of Physical Mechanical Property of Vulcanizate During the Period of Heat Ageing. *China Synth. Rubber Ind.* **1997**, *2*, 44–53. (In Chinese) [CrossRef]

36. Symposium, R.M. Annual Reliability and Maintainability Symposium. In Proceedings of the International Symposium on Product Quality & Integrity, Tampa, FL, USA, 27–30 January 2003.

37. Zhang, X.; Chang, X.; Chen, S.; Fan, S. Hygrothermal aging testing and life assessment for adhesive interface of solid rocket motor. *J. Solid Rocket. Technol.* **2013**, *36*, 27–31. (In Chinese) [CrossRef]

38. GBT 23446-2009; Spray Polyurea Waterproofing Coating. National Standards of the People's Republic of China: Beijing, China, 2021.

Article

The Mechanism of Deformation Compatibility of TA2/Q345 Laminated Metal in Dynamic Testing with Split-Hopkinson Pressure Bar

Yanshu Fu [1], Shoubo Chen [1], Penglong Zhao [1,*] and Xiaojun Ye [1,2]

[1] School of Advanced Manufacturing, Nanchang University, Nanchang 330031, China; yshfu@ncu.edu.cn (Y.F.); 13767030491@163.com (S.C.); yexiaojun@ncu.edu.cn (X.Y.)
[2] School of Information and Artificial Intelligence, Nanchang Institute of Science & Technology, Nanchang 330108, China
* Correspondence: ncuzpl@ncu.edu.cn

Abstract: The laminated metal materials are widely used in military, automobile and aerospace industries, but their dynamic response mechanical behavior needs to be further clarified, especially for materials with joint interface paralleling to the loading direction. The mechanical properties of TA2/Q345 (Titanium/Steel) laminated metal of this structure were studied by using the split Hopkinson pressure bar (SHPB). To shed light on the stress-state of a laminated metal with parallel structure, the relative non-uniformity of internal stress $R(t)$ was analyzed. The mechanism of deformation compatibility of welding interface was discussed in detail. The current experiments demonstrate that in the strain rate range of 931–2250 s^{-1}, the discrepancies of the internal stress in specimens are less than 5%, so the stress-state equilibrium hypothesis is satisfied during the effective loading time. Therefore, it is reasonable to believe that all stress–strain responses of the material are valid and reliable. Furthermore, the four deformation stages, i.e., the elastic stage, the plastic modulus compatible deformation stage, uniform plastic deformation stage and non-uniform plastic deformation stage, of the laminated metal with parallel structure were firstly proposed under the modulating action of the welding interface. The deformation stages are helpful for better utilization of laminated materials.

Keywords: deformation compatibility; stress-state equilibrium; laminated metal

Citation: Fu, Y.; Chen, S.; Zhao, P.; Ye, X. The Mechanism of Deformation Compatibility of TA2/Q345 Laminated Metal in Dynamic Testing with Split-Hopkinson Pressure Bar. *Materials* **2023**, *16*, 7659. https://doi.org/10.3390/ma16247659

Academic Editor: Daniel N. Blaschke

Received: 13 October 2023
Revised: 8 December 2023
Accepted: 12 December 2023
Published: 15 December 2023

1. Introduction

Laminated metal materials are widely used in many applications such as military, automobile and aerospace industries due to a number of unique combinations of high physical, mechanical and operational properties [1–4]. The laminated materials can consist of different kinds of layers such as ceramic/metal, metal/metal, and metal/composite and so on. Studies have shown that the multilayered composite structure is lighter in weight, has higher impact resistance, and more designable than homogeneous materials [5,6]. Fernando et al. [7] presented a comprehensive analysis of the blast response of functionally graded composite metallic plates and observed that the impedance graded composite plates, which were lighter in density than the monolithic plates, resisted the highly intensive blast loads through their enhanced ductility. Gladkovsky et al. [8] researched the microstructure and mechanical properties of sandwich copper/steel composites materials. They pointed out that the composites have higher strength properties than initial cooper by approximately 1.8–3.5 times. Besides the quasi-static compression tests, the dynamic responses are also different between them [9]. As for composites or laminated materials, it is essential to understand their mechanical behavior and constitutive model under wide strain rate loading [10,11]. Zhang et al. [12] employed bumpers constructed from Impedance-graded materials (IGM) to improve meteoroid/debris shielding structures for spacecrafts. In their research works, the wave propagation and thermodynamic states in the bumper materials were discussed, and it was shown that the impedance-graded bumper changes the wave propagation path and duration time, which lead to enough time to make materials very hot and break up, and the solid projectile fragments are expanded over a greater area.

The split Hopkinson pressure bar (SHPB) technique is the most classical method for obtaining response properties of materials at dynamic compressions [13–17]. It was developed for detecting explosive waves by Hopkinson and developed revolutionarily by Kolsky in 1949 for obtaining properties of materials at high strain rates [18,19]. It is worth to note that the dissimilar materials with differing moduli and impedances will cause complex wave reflection and transmission phenomena at each encountered interface [20]. The impedance represents the resistance a material presents to the transmission of stress waves, and it is a product of the material's density and wave propagation speed. The stress wave impedance mismatch refers to the mismatch in the acoustic or mechanical impedance between two materials when a stress wave propagates from one material to another. When a vertical or oblique, but not parallel, stress wave is incident and encounters an interface between two materials with different impedance values, some of the wave's energy can be reflected back and some can be transmitted to the other material. The extent of reflection and transmission depends on the difference in impedance between the materials. If the impedance mismatch is significant, a large portion of the wave's energy can be reflected, leading to poor transmission of stress and potentially causing issues such as energy loss, reduced signal quality, or even structural damage. Hui et al. [1] pointed out that the wave propagation of shock wave's reflection behavior in an IGM bumper has led to better attenuation of shock wave energy by multiple interface reflections and transmissions, which should play an important role in the heating and fracture effect of bumpers and projectiles.

As for laminated materials, the mechanism of deformation compatibility under dynamic impact loading is very important. However, the research works mentioned above only paid attention to the behavior of laminated materials when the impact loaded vertically to the welding interface. They lacked a study of the effect of the impact loading component, which is parallel to the welding interface. Actually, when impact loads are applied to laminated materials, the vertical and parallel components almost always accompany each other, so it should be necessary to detect the behavior of laminated materials responding to the parallel structure of the multi-layer interfaces. There are, as of yet, no experimental, numerical or analytical studies available in the published literature. This present work was initially motivated by a need to address the research gap of the lack of comprehensive studies for multi-layered metal plates responding to the impact loading parallel component of the multi-layer interfaces. As such, experimental and analytical studies were conducted for an impedance mismatch composite metal plate using the SHPB test system and mathematical model.

The remainder of this paper is organized as follows. In Section 2, we describe the configurations of the experimental setup and the specimen preparations in detail. Then, we present a series of analyses concerning the design of the experiments, the mode-mix analysis, and the post-processing of the raw data for extraction of the mixed-mode properties as well as a complete set of results in Section 3, where the effects of the separation rate and mode-mix are discussed at length. Conclusions are provided in Section 4.

2. Material and Methods

2.1. Experimental Setup

A conventional SHPB apparatus is shown in Figure 1, consisting of a gas gun, a projectile or striker bar ($\varphi 14.5 \times 200$ mm), an incident bar ($\varphi 14.5 \times 1000$ mm), a transmitted bar ($\varphi 14.5 \times 1000$ mm), an energy absorber bar ($\varphi 14.5 \times 600$ mm) and a high-frequency data processing system. The specimen was placed between the incident and transmitted bars. As the striker bar hits the incident bar at a constant speed v_0, an incident wave is generated and transmitted throughout the bar. When it propagates to the interface between the incident bar and specimen, the incident wave will reflect and transmit due to the different acoustic impedances of the contacted materials which is called acoustic impedance mismatch hereinafter. Part of the incident wave becomes a reflected wave and transfers towards the incident bar with velocity in the opposite direction, while the other part wave becomes a transmission wave and propagates through the specimen. The amounts of reflected and transmitted waves at the interfaces depend on the mechanical impedance ratio of the bars and the specimen. Soon, at the specimen/transmitted bar interfaces, the reflection and transmission phenomenon happened again. Eventually, the energy of the transmitted wave will be dissipated in the absorber bar.

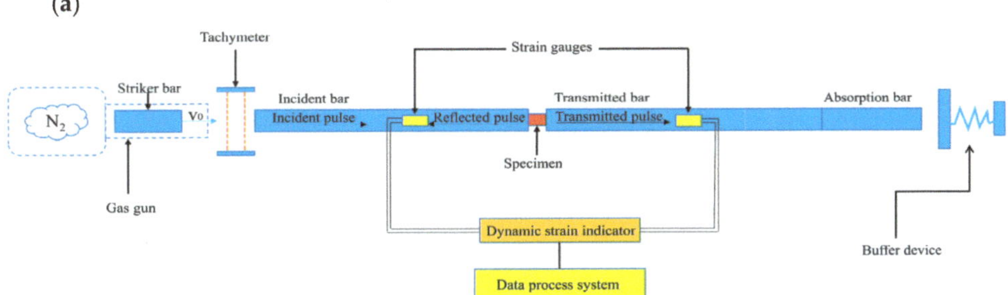

Figure 1. Schematic of SHPB (**a**) schematic diagram, and (**b**) experimental apparatus.

It is necessary to ensure the bars in the apparatus are always keeping linear elasticity during the experiments, and the length of the bar must be far larger than the diameter to ignore the transverse inertial effect, so that the data can be processed based on the one-dimensional stress wave theory. The strain rate $\dot{\varepsilon}$, nominal strain ε, and the nominal stress σ are given as follows:

$$\dot{\varepsilon}(t) = \frac{C_0}{L_s}(\varepsilon_i - \varepsilon_r - \varepsilon_t) \tag{1}$$

$$\varepsilon(t) = \frac{C_0}{L_s}\int_0^t (\varepsilon_i - \varepsilon_r - \varepsilon_t)dt \tag{2}$$

$$\sigma(t) = \frac{A}{2A_s}E(\varepsilon_i + \varepsilon_r + \varepsilon_t) \tag{3}$$

where C_0, A and E are the wave speed, cross-section area and the elastic modulus of the bar, respectively. A_s and L_s are the initial cross-section area and length of the specimen. ε_i, ε_r and ε_t are the incident strain wave, the reflected strain wave and the transmitted strain wave, respectively.

Assuming that the specimen deforms uniformly in longitudinal direction, then

$$\varepsilon_i + \varepsilon_r = \varepsilon_t \tag{4}$$

Substituting Equation (4) into Equations (1)–(3), we obtain:

$$\dot{\varepsilon}(t) = -\frac{2C_0}{L_s}(\varepsilon_i - \varepsilon_t) = -\frac{2C_0}{L_s}\varepsilon_r \tag{5}$$

$$\varepsilon(t) = -\frac{2C_0}{L_s}\int_0^t (\varepsilon_i - \varepsilon_t)dt = -\frac{2C_0}{L_s}\int_0^t (\varepsilon_r)dt \tag{6}$$

$$\sigma(t) = \frac{A}{A_s}E(\varepsilon_i + \varepsilon_r) = \frac{AE}{A_s}\varepsilon_t \tag{7}$$

By using Equations (5)–(7), the stress–strain curves corresponding to series of strain rates can be deduced.

In this study, the stress signals were measured by the strain gauges that attached on the incident and reflected bars. The sensitivity of the gauge is 1 V/1000 με, and the range is 0~±3 × 10^4 με. The movement of the striker is controlled by high-pressure Nitrogen, with an adjustment accuracy of 0.01 MPa. The striker bar velocities are 6.31 m/s, 13.63 m/s, 16.73 m/s, 19.24 m/s, and 22.03 m/s, denoted as cases 1–5 hereafter, as shown in Table 1.

Table 1. Emission pressure and the velocity of striker bar.

Cases	1	2	3	4	5
Velocity of striker bar (m/s)	6.31	13.63	16.73	19.24	22.03
Emission pressure (MPa)	0.22	0.25	0.35	0.48	0.60

2.2. Specimen Preparations

The laminated metal is fabricated using explosive welding of 30 mm-thick base steel material Q345 and 8 mm thick composite titanium material TA2. Explosive welding, belonging to solid-state welding, has been used to connect a wide range of dissimilar impedance materials and obtain firm bonds through high pressure and heat without causing significant crystallization or phase transition [21]. The metal performs act as an elastic-viscous plastic fluid under the detonation wave. The type of explosive is rock ammonium nitrate, with a charge density of 0.8 g/cm^3. The explosive detonates at a speed of 2800 m/s, and the gap between the basic plate (Q345) and the fly plate (TA2) is about 6 mm. Under the proper welding parameters, the repeatability of the laminated materials could be ensured.

Different from the pure Q345 and TA2, the properties of these two metals are changed after explosive welding. The elasticity modulus of Q345 and TA2 in laminated materials is tested with nano-indentation of iMicor (shown in Figure 2). The tests points were arranged from TA2 to Q345 at 210 μm from the welding interface with a spacing of 50 μm. The elasticity modulus is averaged with 5 test points, respectively. The properties of Q345 and TA2 as well as the SHPB bars are listed in Table 2. The microstructure of TA2/Q345 welding interface is shown in Figure 3. It can be seen that there exists a wavy interface at the steel and titanium boundary with no obvious defects. The wavy interfaces ensure that the laminated metal has favorable mechanical properties and a strong bond area. It is clear from Figure 3 that twice the amplitude and half the wavelength are 0.25 mm and 0.61 mm, respectively. Since the wavelength is almost ten times the amplitude, the welding interface can be regarded as a plane macroscopically for the sake of simplicity.

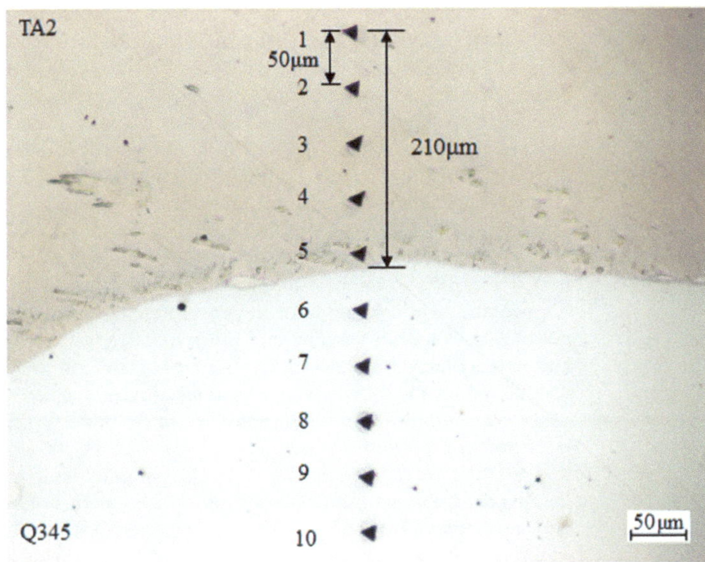

Figure 2. Nano-indentation test of TA2/Q345 laminated metal (The numbers 1 to 10 correspond to ten test points).

Table 2. Materials' parameters of TA2, Q345 and SHPB bars.

	Density (g/cm³)	Hardness HV	Elasticity Modulus (GPa)	Elastic Wave Velocity (m/s)	Poisson Ratio
Q345	7.83	≥160	246.18	5607.2	0.3
TA2	4.51	≥140	132.82	5426.8	0.33
SHPB bars	7.69	≥500	200	5100.0	0.3

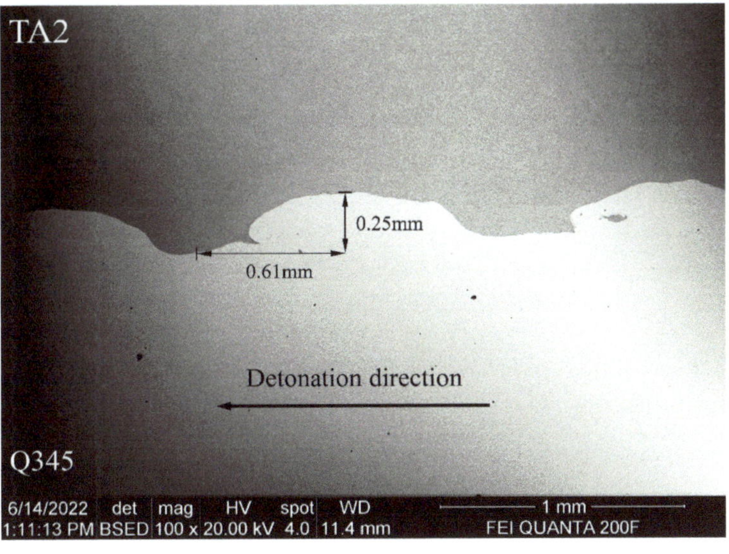

Figure 3. Microstructure of TA2 and Q345 welding interface.

As previously mentioned, the initial assumption in the wave analysis of the SHPB is the stress-state equilibrium in the specimen, meaning that the force on the incident bar side of the specimen is equal to that on the transmitted bar side of the specimen [22–25]. In order to achieve the required stress-state equilibrium and stress level in the specimen, the length of the specimen should be greater than the transit time for the stress pulse in the specimen, and the cross-section area of the specimen should be kept smaller than the bar. For trial and error, the specimen is designed a cylinder with a diameter of 8 mm and a length of 4 mm. The geometrical parameter of the specimen and the sampling method are shown in Figure 4a. The original welding TA2/Q345 laminated metal is about 400 mm × 300 m × 38 mm. In order to process the specimen, a cube of 30 mm × 50 mm × 38 mm material is cut off first. Then, a $\phi 8 \times 4$ mm cylinder was made through wire electrical discharge technology (Figure 4b). The two ends of the specimen are polished with sanding to ensure their parallelism is within the tolerance range of 0.01 mm. The weight of the specimen is 1.24 g, and it was placed between the incident and transmitted bars with Vaseline. Vaseline was used to reduce the friction on both end faces of the specimen and to help keep the specimen contacting with the bars at the center line of the bars. As can be seen, the welding interface is located in the middle, and the volume fraction of Q345 and TA2 accounts for 50% each. In this research, we define samples as parallel samples, meaning that the interface parallels the loading direction, so the stress wave propagates along the interface. Because of the physical and mechanical properties differences between Q345 steel and Ti, the wave impedance is different from each other, but they were combined together by explosive welded technology, so samples in this work are made from impedance-mismatched multilayered materials. (All devices, instruments, and materials are used or made in Nanchang City, Jiangxi Province, China).

(a)

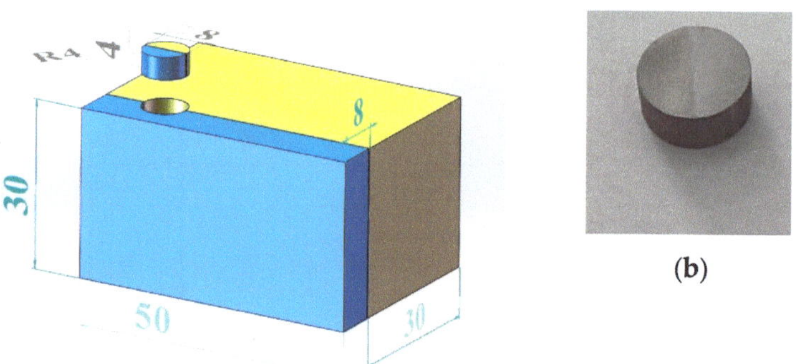

(b)

Figure 4. Geometrical parameter of the TA2/Q345 and specimen (**a**) and specimen diagram (**b**).

3. Results and Discussion

3.1. Typical Waveform of SHPB Tests of TA2/Q345

In the experiments, the stress wave propagation characteristics of the parallel specimen were obtained at different impact loads controlled by the velocity of the striker bar. Each case was repeated three times, and the stress-time as well as the strain rate–time curves of the specimens are shown in Figure 5. The averaged values of the repeated experiments corresponding to the same case were presented with an error bar to show the statistical analyses of the achieved results.

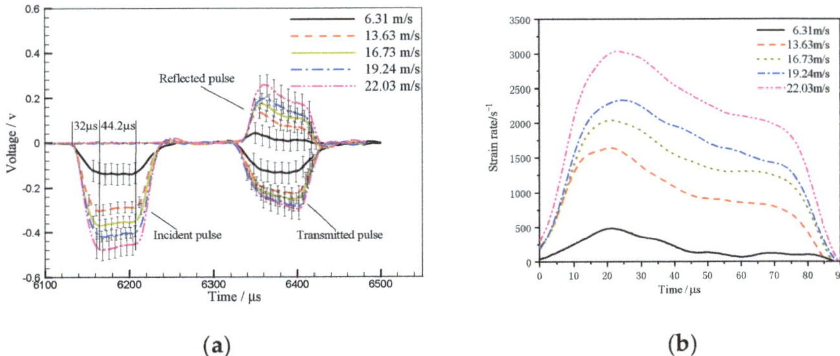

Figure 5. The stress-time curves (**a**) and stress rate–time curves (**b**) of cases 1–5.

As shown in Figure 5a, the rise time of an incident wave is ~32 µs and the corresponding platform segment after reaching the maximum value lasts about 44.2 µs. As for the reflected wave, it declines gradually over time after rising to the peak, indicating the high-work-hardening rate of TA2/Q345 laminated metal materials [26]. Since the shape and amplitude of the transmitted wave are determined by the specimen's stress–strain behavior, the gradual increase of the transmitted wave's amplitude in Figure 5a indicates the strain-hardening response of TA2/Q345.

The strain rate can be calculated from the reflected pulse according to Equation (5) and the strain rate–time curves of cases 1–5 are plotted in Figure 5b. The strain-rate is not constant during the loading process as the reflected wave does not display constant amplitude in Figure 5a. This phenomenon is closely related to the material characteristics of TA2/Q345. The effective loading duration of incident wave is 32–76.2 µs. Therefore, the average strain rate of the stage is regarded as the mean strain rate in the whole loading process. In this study, the strain rates corresponding to cases 1–5 are 146 s^{-1}, 931 s^{-1}, 1384 s^{-1}, 1686 s^{-1} and 2250 s^{-1}.

3.2. Verification of No-Slip Condition at Welding Interface

During the experiment, the specimen was deformed under the action of dynamic compression load. In the deformation process, Q345 and TA2 have the same strain, i.e., no slip at the interface during the deformation process. No slip at the interface means that within a certain strain rate range, the welding interface will not be damaged; Q345 and TA2 are bound tightly, and Q345 and TA2 are co-deformed under the action of the welding interface. In order to verify whether the condition of no slip at the interface during the impact is established, the specimen is cut axially after impact, and the state of the welding interface is observed under Axioscope optical microscope after grinding and polishing to investigate whether slip damage occurs.

Figure 6 shows the microstructure of the welded interface of case 4. It can be seen from Figure 6 that the bonding quality of the welded interface is still good at this strain rate. The phenomenon observed under dynamic loading conditions with strain rate lower than 1686 s^{-1} is consistent with Figure 6. Therefore, it can be inferred that the condition of no slip at the interface is valid in the range of strain rate from 0 to 1686 s^{-1}. The damaged interface of case 5 is shown in Figure 9, which will be discussed in detail later.

In addition, the axial plastic deformation compression of the specimen is 0.49 mm of case 4. This indicates that the welding interface failure must occur after the specimen has undergone plastic deformation, and it can be inferred that the bonding strength of the welding interface is greater than the yield strength of Q345 and TA2, respectively.

Figure 6. The microstructure of the welding interface of case 4.

3.3. Verification of the Specimen Stress Uniformity

According to the data processing principle of SHPB, it can be seen that the stress balance at both ends of the specimen in the experiment is sufficient and necessary to ensure the real and reliable experimental results. Therefore, it is very important to verify whether the specimen achieves stress uniformity quickly during loading. At present, it is generally accepted that when the stress wave is reflected back and forth in the specimen for two or three times, if the difference between the stress at both ends of the specimen and the ratio of its average value (in this case, the average value of the stress at both ends of the specimen is used as the average stress of the specimen) is less than 5%, the uniformity hypothesis can be considered satisfied [27]. However, because the material used in this paper is a TA2/Q345 laminate material, there is a large difference in material properties between the substrate (Q345) and the composite (TA2). Therefore, it is necessary to examine the stress state at both ends of the specimen during the experiment and evaluate the degree of stress uniformity inside the specimen to judge whether the experiment meets the uniformity hypothesis. According to the one-dimensional stress wave theory:

$$\sigma_t^L = \frac{EA}{A_s}(\varepsilon_i + \varepsilon_r)$$

(8)

$$\sigma_t^R = \frac{EA}{A_s}\varepsilon_t$$

(9)

where σ_t^L is the stress at the left end of the specimen at time t (the end of the specimen in contact with the incident bar is the left end) and σ_t^R is the stress at the right end of the specimen at the same time (the end of the specimen in contact with the transmission bar is the right end). Therefore, the stress difference between the two ends of the specimen $\Delta\sigma_t$ and the average stress of the specimen $\overline{\sigma_t}$ at time t are:

$$\Delta\sigma_t = \sigma_t^L - \sigma_t^R$$

(10)

$$\overline{\sigma_t} = \frac{\sigma_t^L + \sigma_t^R}{2}$$

(11)

The relative non-uniformity of the internal stress of the specimen at time t is defined as $R(t)$ [28]:

$$R(t) = \left| \frac{\Delta \sigma_t}{\overline{\sigma_t}} \right| \tag{12}$$

By substituting Equations (8)–(11) into Equation (12), we get:

$$R(t) = \left| \frac{2(\varepsilon_I + \varepsilon_R - \varepsilon_T)}{\varepsilon_I + \varepsilon_R + \varepsilon_T} \right| \tag{13}$$

According to the existing evaluation criteria, if $R(t) < 5\%$ during the effective loading time of incident wave (incident wave platform segment), the experiment can be considered to meet the uniformity hypothesis and the experimental results are valid.

In this paper, the variation of $R(t)$ with time under different strain rate loading conditions is shown in Figure 7. As can be seen from Figure 7, for cases 2–5, $R(t)$ of the effective loading time of the incident wave (as can be seen from Figure 5a that the effective loading time of the incident wave ranges from 32 μs to 76.2 μs) is less than 5%. Therefore, the experimental results are consistent with the assumption of uniformity, the experimental scheme is reasonable, and the experimental results are valid. However, as for case 1, $R(t)$ is less than 5% when it is around 50 μs, and the uniformity hypothesis is satisfied only in the latter half of the effective loading time of the incident wave. Therefore, the stress–strain relationship derived from the reflected and transmitted waves measured for case 1 is not accurate. The material properties of TA2/Q345 laminates at this strain rate cannot be reflected correctly.

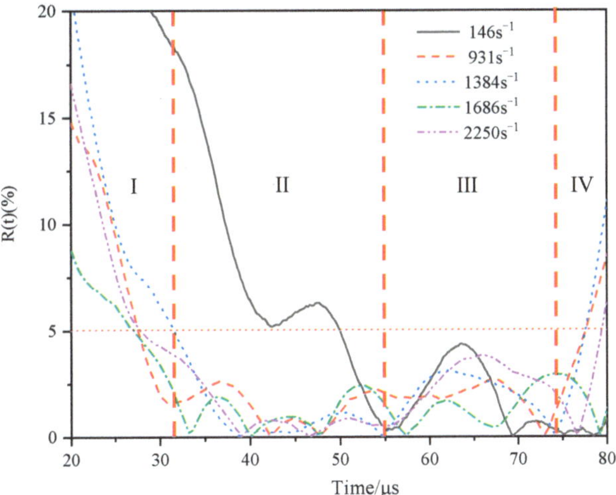

Figure 7. The variation of $R(t)$ with time of cases 1–5.

As shown in Figure 7, during the experiment, the changes of $R(t)$ corresponding to cases 2–5 showed four different characteristics with time in turn. The $R(t) - t$ curve can therefore be divided into four regions. They are I zone, II Zone, III zone and IV zone. In zone I, $R(t)$ is always larger than 5%, which indicates that the stress state in the specimen during this process is not uniform. In zone II, $R(t)$ is always less than 5% (except for $R(t)$ in case 1), and remains around 2% after 40 μs. It indicates that the internal stress uniformity of the specimen remains at a high level during this period of time.

After entering the III zone, although $R(t)$ is still less than 5%, significant fluctuation occurs, and the stress uniformity becomes worse. It is shown that the compatibility ability of the welding interface becomes worse under continuous stress wave loading. The fluctuation phenomenon appeared in cases 2–5, but only the initial time of the fluctuation was slightly different. There may be two reasons for the fluctuation: one is that the welding interface is damaged at this stage, resulting in the non-slip condition of the interface no longer being satisfied; the second reason is that the transverse deformation of the specimen accumulates to a certain extent at this stage, which intensifies the transverse inertia effect. According to Section 3.2, under the loading condition of strain rate from 0 to 1686 s^{-1}, the welding interface is still valid, indicating that the fluctuation is mainly caused by the transverse inertia effect. In region III, $R(t)$ is always less than 5%, indicating that the transverse inertia effect is not obvious. It also shows that the specimen satisfies the one-dimensional stress hypothesis during loading. The starting time of the data fluctuation in $R(t)$ in region IV is around 75 µs. Combined with Figure 5a, it can be seen that the incident wave begins to unload at 76.2 µs. Therefore, the reason for the fluctuation of data in region IV is the unloading of the incident wave.

3.4. Four Deformation Stages during the Dynamic Response Process

According to the analysis of various regions of the $R(t) - t$ curve in Figure 7 in the previous section, combined with the one-dimensional stress wave theory, it can be inferred that TA2/Q345 laminated specimens will enter the four deformation stages successively under the action of dynamic compression load under the mechanism of coordinated deformation participation of the welding interface. These are: the respectively elastic deformation stage, the plastic modulus compatibility deformation stage, the uniform plastic deformation stage and the non-uniform plastic deformation stage.

3.4.1. Elastic Deformation Stage

As shown in Figure 8, A_0, A_1, and A_2 are the cross-sectional areas of the incident bar, TA2 and Q345, respectively. A_3 is the area of the welding interface. According to Newton's third law, the initial stress state of the contact surface between the incident bar and the specimen is:

$$\sigma_b A_0 = \rho_1 c_1 v_1 A_1 + \rho_2 c_2 v_2 A_2 \tag{14}$$

$$\sigma_b = E\varepsilon_i \tag{15}$$

where, $\rho_1 c_1$ and $\rho_2 c_2$ are the wave impedance of TA2 and Q345, ρ_1 is the density of TA2, ρ_2 is the density of Q345, c_1 is the elastic wave velocity of TA2, c_2 is the elastic wave velocity of Q345, v_1 is the material point velocity of TA2 side, v_2 is the material point velocity of Q345 side, σ_b is the stress of the incident bar, ε_i is the strain of the incident bar. Due to the existence of the welding interface, as long as the interface does not break, the interface no slip condition is met:

$$v_1 = v_2 = v_3 \tag{16}$$

where v_3 is the material point velocity on the welding interface, substituting Equations (15) and (16) into Equation (14):

$$v_3 = \frac{E\varepsilon_i A_0}{\rho_1 c_1 A_1 + \rho_2 c_2 A_2} \tag{17}$$

In the elastic stage, the internal stress σ_1 of TA2 side and σ_2 of Q345 side are:

$$\sigma_1 = \rho_1 c_1 v_1 \tag{18}$$

$$\sigma_1 = \rho_1 c_1 v_1 \tag{19}$$

Because of the wave impedance mismatch of TA2 and Q345, it must be impossible for the specimen to reach stress equilibrium in the elastic deformation stage, and there is shear stress at the interface:

$$\sigma^e{}_s = \sigma_1 - \sigma_2 = \frac{(\rho_1 c_1 - \rho_2 c_2) E\varepsilon_i^e A_0}{\rho_1 c_1 A_1 + \rho_2 c_2 A_2} \tag{20}$$

where ε_i^e is the strain corresponding to the elastic deformation stage in the strain-time curve of the incident bar.

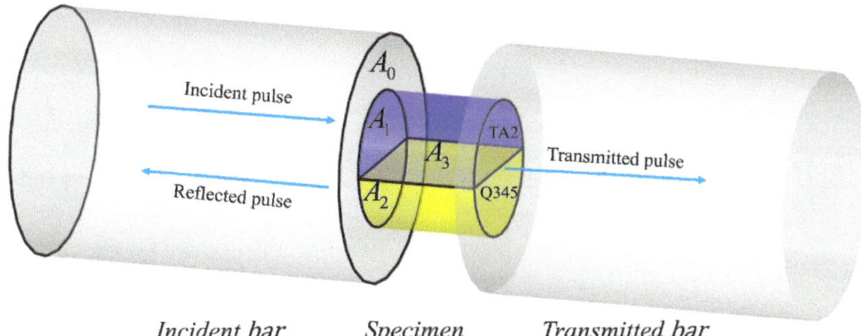

Figure 8. Loading diagram of TA2/Q345parallel specimen.

In the elastic deformation stage, the strain is uniform but the stress is not. As can be seen from Table 3, the time of elastic deformation stage of the specimen ranges from 0 to 6.2 µs, while the starting time of zone I is about 20 µs. This stage occurs before zone I in the $R(t) - t$ diagram.

Table 3. The key node information in the true stress–strain relationship of cases 2–5.

Strain Rate (s^{-1})	Yield Strength (Mpa)	Yield Strain (10^{-3})	Yield Time (µs)	Initial Stress of Strain Hardening Effect (Mpa)	Initial Time of Linear Hardening (µs)
931	141.26	2.98	6.2	597.92	29.2
1384	169.47	2.56	5.8	638.56	33.2
1686	179.78	2.95	6	673.37	32
2250	204.87	4.22	6	695.33	34.2

3.4.2. Plastic Modulus Compatibility Deformation Stage

After the elastic deformation stage, the specimen entered the Plastic modulus compatibility deformation stage under the action of dynamic compression load. There may be three reasons for $R(t)$ greater than 5% in zone I: First, it takes time for the stress wave to propagate in the specimen, and the stresses at both ends of the specimen are not equal in the initial response stage. The second reason is that the wave impedances of Q345 and TA2 are different. Under the condition of no slip at the interface, the stress distributions inside Q345 and TA2 are not uniform when the elastic deformation occurs. The axial stress distributions were found to be non-uniform in the elastic deformation range of the specimen. The third reason is that after the specimen enters the plastic deformation, it takes time for the welding interface to coordinate the internal stress of Q345 and TA2 from unequal to equal. Therefore, in the initial stage of plastic deformation, the stress at both ends of the specimen is unequal, but the plastic deformation is conducive to uniform stress distribution and transverse stress balance, while the plastic deformation is a tendency to produce a more homogeneous stress distribution within the components. The performance of $R(t)$ in zone I can be deduced as the third one. Therefore, when the dynamic loading strain rate is low (such as in case 1), the specimen cannot quickly enter the plastic deformation stage, and $R(t)$ cannot reach 5% at the end of zone I.

3.4.3. Uniform Plastic Deformation Stage

According to regions II and III in Figure 7, the TA2/Q345 laminated specimen had a uniform stress deformation stage during the deformation process. Therefore, assuming that the internal stress on TA2 side is equal to that on Q345 side at this stage, it must be satisfied:

$$\rho_1 c_{p1} = \rho_2 c_{p2} \tag{21}$$

where c_{p1} and c_{p2} are the plastic wave velocity of the titanium and the steel, respectively. According to the stress wave theory:

$$c_p = \sqrt{\frac{E_p}{\rho}} \tag{22}$$

$$E_p = \frac{d\sigma_p}{d\varepsilon_p} \tag{23}$$

where c_p is the plastic wave velocity of the material, ρ is the density of the material, σ_p is the plastic stress of the material, ε_p is the plastic strain of the material and E_p is the plastic modulus of the material. From Equations (21)–(23):

$$\rho_1 E_{p1} = \rho_2 E_{p2} \tag{24}$$

where E_{p1} and E_{p2} are the plastic modulus of titanium and steel, respectively.

According to Equation (24), in this deformation stage, the welding interface plays a role in adjusting the plastic modulus of TA2 and Q345, so that the plastic stresses on both sides of the welding interface are equal, the specimens are uniformly deformed under the dynamic compression load, and the shear stress of the welding interface is zero at this stage. At this stage, the strain is uniform, and the stress is balanced during the deformation of the specimen. Therefore, this stage is called the uniform plastic deformation stage, corresponding to regions II and III in Figure 7.

After the elastic deformation stage and before the uniform plastic deformation stage, due to the compatibility of the plastic moduli of TA2 and Q345 at the welding interface, it takes time for the plastic stress on both sides of the welding interface to be unequal to equal. At this time, $R(t)$ gradually approaches 5%, from greater than 5% to less than 5%. This compatibility process is called the plastic modulus compatibility deformation stage and corresponds to zone I in Figure 7.

3.4.4. Non-Uniform Plastic Deformation Stage

In case 5, when the strain rate increases to 2250 s^{-1}, the axial plastic deformation of the specimen is 0.62 mm. At this time, the microscopic state of the welding interface is shown in Figure 9. It can be seen that the specimen cracks along the welding interface, which indicates that when the loaded strain rate reaches a certain value, the shear stress at the welding interface exceeds the compatibility capacity and fails. In the process of load response, the specimen will enter the non-uniform deformation and plasticity stage after undergoing uniform deformation and plasticity stage. At this time, the strain is uniform and the stress is not uniform in the deformation process of the specimen. Shear stress occurs again at the interface. When the interface shear stress is greater than the bonding strength of the welding interface, the welding interface will be damaged. After the welding interface is damaged, the condition of no slip is no longer satisfied, and Q345 and TA2 are deformed, respectively, under the loading of stress waves. At this time, the strain and stress of the specimen are not uniform during deformation.

Figure 9. The microstructure of the welding interface of case 5.

Since the non-uniform deformation plastic stage occurs only when the loading strain rate is greater than a certain value and the time when this stage occurs is outside the range of experimental measurement (after 76.2 μs), the stress non-uniformity at this stage does not affect the stress–strain relationship obtained in the end.

3.5. Dynamic Compressive Mechanical Response of TA2/Q345

According to Equations (6) and (7), the nominal stress–strain curves of specimens under different strain rates can be obtained, and the nominal stress–strain curves can be converted into true stress–strain curves by simple conversion. The true stress–strain curves of specimens under different strain rates are shown in Figure 10. It can be seen from Figure 10 that the true stress–strain relationship under the loading conditions of four strain rates of cases 2–5 presents a consistent change tendency. In order to further analyze the dynamic compressive mechanical behavior of the specimen, the true stress–strain curve and the corresponding true strain–time curve of case 5 were plotted separately for analysis. The results are shown in Figure 11a and 11b, respectively.

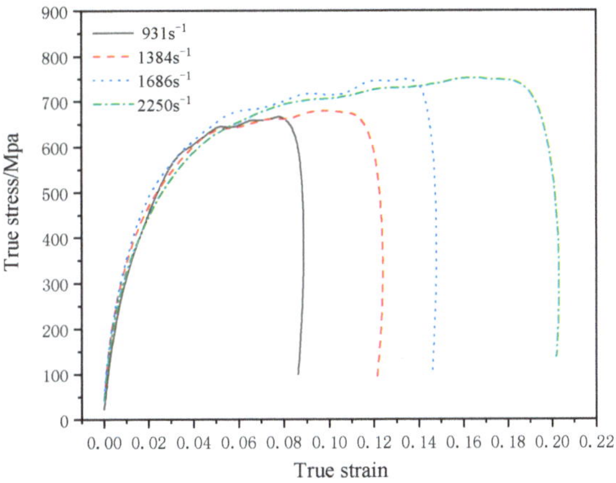

Figure 10. The true stress–strain curves of cases 2–5.

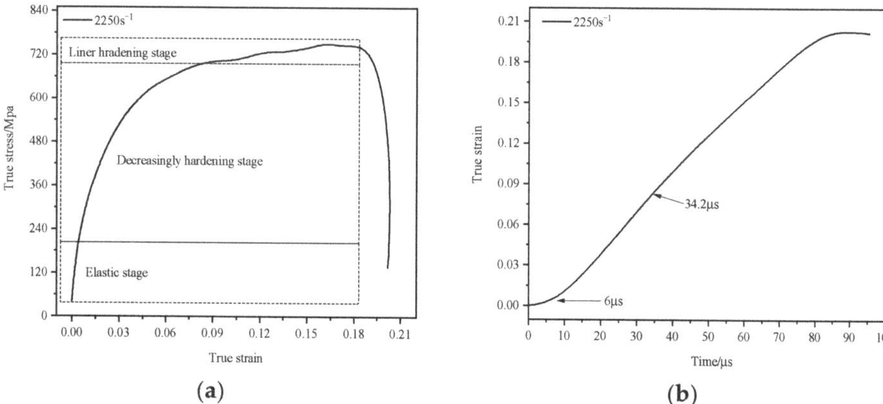

Figure 11. The true stress–strain curve (**a**) and the true strain–time curve (**b**) of case 5.

As can be seen from Figure 11a, the true stress–strain curve can be divided into three stages. These are the elastic stage, the decreasing hardening stage, and the linear hardening stage, respectively. The initial deformation stage of the specimen is elastic. When the stress reaches 204.87 Mpa, the specimen begins to yield. Combined with the true strain–time curve (Figure 11b), the corresponding yield moment is 6. The specimen enters the plastic deformation stage after yielding. At the initial stage of plastic deformation, the plastic modulus of the specimen presents a decreasing hardening phenomenon, and at the late stage of plastic change, it presents a linear hardening phenomenon. The starting time of the linear hardening stage is 34.2 µs, corresponding to the starting time of zone II, indicating that the decreasing hardening stage is the plastic modulus compatibility stage, corresponding to zone I of the $R(t) - t$ curve, and the linear hardening stage corresponds to zone II and zone III, indicating that the linear hardening stage is the uniform plastic deformation stage. The true stress–strain response of the specimen is consistent with the theoretical analysis of specimen deformation in Section 3.3.

In order to further analyze the dynamic compressive mechanical response behavior of TA2/Q345 laminated materials under parallel structure, the key node information in the true stress–strain curve under different strain rates was sorted into a table, and the results were shown in Table 3.

As can be seen from Table 3, with the increase of strain rate from 931 s^{-1} to 2250 s^{-1}, the yield strength of TA2/Q345 laminates increases from 141.26 Mpa to 204.87 Mpa. It can be seen that TA2/Q345 laminated materials have a certain strain rate strengthening effect. TA2/Q345 laminated materials can be considered as strain rate-sensitive materials in the strain rate range of 931 s^{-1}–2250 s^{-1}. According to Figure 11, in the elastic stage, the stress increases sharply with the increase of strain, indicating that TA2/Q345 laminated materials have obvious strain hardening effect. With the increasing strain rate, the plastic strain of TA2/Q345 laminates increases significantly. At 931 s^{-1}, the plastic strain is 0.08, and at 2250 s^{-1}, the plastic strain increases to 0.18. The plastic fluidity of TA2/Q345 laminates is significantly enhanced, reflecting the obvious strain rate plasticizing effect. The reason for this strain rate plasticizing effect is that the specimen is in an adiabatic state during the deformation process of high strain rate, and the thermal energy generated by the impact load causes the temperature of the specimen to rise rapidly, softening the specimen and thus increasing the plastic fluidity of the specimen [29,30]. In the linear hardening stage, the true stress–strain curve presents periodic fluctuations, and strain hardening and strain softening alternately occur. This phenomenon may be caused by the competition between the strain hardening effect and the specimen softening effect caused by the adiabatic temperature rise [31].

4. Conclusions

The mechanism of deformation compatibility of parallel structure TA2/Q345 laminated materials is investigated using a 14.5 mm diameter SHPB system with strain rates in the range of 146 s^{-1}~2250 s^{-1}. The laminated material is fabricated using the explosive welding method. The welding interface was analyzed in detail before and after different dynamic compression loads. By analyzing the stress evolution law, the following conclusions are obtained:

1. The relative non-uniformity of the internal stress is lower than 5% when the strain rates are in range of 931–2250 s^{-1}, indicating that the stress–strain relationships of the parallel structure specimens are reliable. The experimental results are real and effective. When the strain rate is at a relatively low level, i.e., strain rate is 146 s^{-1}, the one-dimensional stress wave hypothesis and the stress uniformity hypothesis are not satisfied until the latter half of the effective loading time of the incident wave.
2. There exist four deformation stages of the parallel structure TA2/Q345 laminated material under the dynamic compression loading condition, namely, the elastic deformation stage, the plastic modulus compatible deformation stage, the uniform plastic deformation stage and the non-uniform plastic deformation stage. During the whole loading process, the proportion of the elastic deformation stage is only 7.5%. The plastic modulus compatible deformation stage accounts for 15.0%, mainly depending on the stress wave propagating in the specimen. The uniform plastic deformation stage accounts for the highest proportion of 53.7%, which is the primary stage in the process. The non-uniform plastic deformation stage happened with obvious characteristic of failure welding interface.
3. The plastic modulus-compatible deformation stage is characterized by the decreasing hardening phenomenon, while the uniform plastic deformation stage is characterized by the linear hardening of the plastic modulus. In the non-uniform plastic deformation stage, adiabatic temperature dominates the material behavior, leading to the specimen softening effect.
4. The parallel structure TA2/Q345 composites exhibit strain rate hardening effect, strain rate strengthening effect and strain rate plasticizing effect under 931 s^{-1}–2250 s^{-1} strain rate dynamic compression load, which can be utilized in protective structure in electronic packaging, vehicle collision avoidance system et al.

Future Scope

The primary goal of the current research is to figure out the mechanism of deformation compatibility of laminated metal with a parallel structure. However, the current study's findings are restricted to one-dimensional conditions. Consequently, more research in three-dimensions will be needed in the future on the use of laminated metal in engineering.

Author Contributions: Formal analysis, Y.F.; Investigation, S.C.; Writing—original draft, X.Y.; Writing—review & editing, P.Z. All authors have read and agreed to the published version of the manuscript.

Funding: This research was funded by the National Natural Science Foundation of China (Grant Nos. 12162024, 12062013), the Foundation of Jiangxi Province for Distinguished Young Scholars, China (Grant No. 20192BCB23003), and the Natural Science Foundation of Jiangxi Province, China (Grant No. 20224BAB201019).

Institutional Review Board Statement: Not applicable.

Informed Consent Statement: Not applicable.

Data Availability Statement: Data are contained within the article.

Conflicts of Interest: The authors declare no conflict of interest.

References

1. Hui, D.; Dutta, P.K. A new concept of shock mitigation by impedance-graded materials. *Compos. Part B-Eng.* **2011**, *42*, 2181–2184. [CrossRef]
2. Corigliano, P.; Crupi, V.; Guglielmino, E.; Sili, A.M. Full-field analysis of AL/FE explosive welded joints for shipbuilding applications. *Mar. Struct.* **2018**, *57*, 207–218. [CrossRef]
3. Sudha, J.; Vasumathi, M.; Archiana, P.; Balakumar, M.; Jaidah, M. Low velocity impact and axial loading response of hybrid fibre aluminium laminates. *J. Mech. Sci. Technol.* **2021**, *35*, 4925–4930. [CrossRef]
4. Boggarapu, V.; Gujjala, R.; Ojha, S.; Acharya, S.; Venkateswara Babu, P.; Chowdary, S.; Gara, D.K. State of the art in functionally graded materials. *Compos. Struct.* **2021**, *262*, 113596. [CrossRef]
5. Bruck Hugh, A. A one-dimensional model for designing functionally graded materials to manage stress waves. *Int. J. Solids Struct.* **2000**, *37*, 6383–6395. [CrossRef]
6. Tasdemirci, A.; Hall, I.W. The effects of plastic deformation on stress wave propagation in multi-layer materials. *Int. J. Impact Eng.* **2007**, *34*, 1797–1813. [CrossRef]
7. Fernando, P.L.N.; Mohotti, D.; Remennikov, A. Behaviour of explosively welded impedance-graded multi-metal composite plates under near-field blast loads. *Int. J. Mech. Sci.* **2019**, *163*, 105124. [CrossRef]
8. Gladkovsky, S.V.; Kuteneva, S.V.; Sergeev, S.N. Microstructure and mechanical properties of sandwich cooper/steel composites produced by explosive welding. *Mater. Charact.* **2019**, *154*, 294–303. [CrossRef]
9. Markovsky, P.E.; Janiszewski, J.; Savvakin, D.G.; Stasiuk, O.O.; Cieplak, K.; Baranowski, P.; Prikhodko, S.V. Mechanical behavior of bilayer structures of Ti64 alloy and its composites with TiC or TiB under quasi-static and dynamic compression. *Mater. Des.* **2022**, *223*, 111205. [CrossRef]
10. Zemani, K.; May, A.; Gilson, L.; Tria, D. Numerical investigation of the dynamic behavior of a Ti/TiB functionally graded material using Split Hopkinson Pressure Bar test. *J. Compos. Mater.* **2022**, *57*, 1053–1070. [CrossRef]
11. Samal, M.K.; Sharma, S. A new procedure to evaluate parameters of Johnson-Cook Elastic-Plastic material model from varying strain rate Split Hopkinson Pressure Bar tests. *J. Mater. Eng. Perform.* **2021**, *30*, 8500–8514. [CrossRef]
12. Zhang, P.L.; Gong, Z.Z.; Tian, D.B.; Song, G.M.; Wu, Q.; Cao, Y.; Xu, K.; Li, M. Comparison of shielding performance of Al/Mg impedance-graded-material enhanced and aluminum Whipple shields. *Int. J. Impact Eng.* **2019**, *126*, 101–108. [CrossRef]
13. Davies, E.; Hunter, S. The dynamic compression testing of solids by the method of the split Hopkinson pressure bar. *J. Mech. Phys. Solids* **1963**, *11*, 155–179. [CrossRef]
14. Jankowiak, T.; Rusinek, A.; Voyiadjis, G. Modeling and design of SHPB to characterize brittle materials under compression for high strain rates. *Materials* **2020**, *13*, 2191. [CrossRef]
15. Panowicz, R.; Konarzewski, M. Infulence of imperfect position of a striker and input bar on wave propagation in a split Hopkinson pressure bar setup with a pulse-shape technique. *Appl. Sci.* **2020**, *10*, 2423. [CrossRef]
16. Jang, T.; Yoon, J.; Kim, J. Determination of Johnson-Cook constitutive model coefficients considering initial gap between contact faces in SHPB test. *J. Mater. Res. Technol.* **2023**, *24*, 7242–7257. [CrossRef]
17. Challita, G.; Othman, R. Finite-element analysis of SHPB tests on double-lap adhesive joints. *Int. J. Adhes. Adhes.* **2010**, *30*, 236–244. [CrossRef]
18. Hopkinson, B. A method of measuring the pressure in the deformation of high explosives or by the impact of bullets. *Philos. Trans. R. Soc.* **1914**, *213*, 437–452.
19. Kolsky, H. An investigation of the mechanical properties of materials at very high rates of loading. *Proc. Phys. Soc. B* **1949**, *62*, 676–700. [CrossRef]
20. Tasdemirci, A.; Hall, I.W.; Gama, B.A. Stress wave propagation effects in two and three-layered composite materials. *J. Compos. Mater.* **2004**, *38*, 995–1009. [CrossRef]
21. Guo, Y.; Chen, P.; Arab, A.; Zhou, Q.; Mahmood, Y. High strain rate deformation of explosion-welded Ti6Al4V/pure titanium. *Def. Technol.* **2020**, *16*, 678–688. [CrossRef]
22. Wang, W.; Yang, J.; Deng, G.; Chen, X. Theoretical analysis of stress equilibrium of linear hardening plastic specimen during SHPB tests. *Exp. Mech.* **2023**, *63*, 1353–1369. [CrossRef]
23. Bendarma, A.; Jankowiak, T.; Rusinek, A.; Lodygowski, T.; Jia, B.; Miguélez, M.H.; Klosak, M. Dynamic behavior of Aluminum Alloy Aw 5005 undergoing interfacial friction and specimen configuration in Split Hopkinson pressure bar system at high strain rates and temperatures. *Materials* **2020**, *13*, 4614. [CrossRef] [PubMed]
24. Lifshitz, J.M.; Leber, H. Data processing in the split Hopkinson pressure bar tests. *Int. J. Impact Eng.* **1994**, *15*, 723–733. [CrossRef]
25. Negger, J.; Hoefnagels, J.; Van Der Sluis, O.; Geers, M. Multi-scale experimental analysis of rate dependent metal-elastomer interface mechanics. *J. Mech. Phys. Solids* **2015**, *80*, 26–36. [CrossRef]
26. Vecchio, K.S.; Jiang, F. Improved Pulse Shaping to achieve constant strain rate and stress equilibrium in Split-Hopkinson pressure bar testing. *Metall. Mater. Trans. A* **2007**, *38*, 2655–2665. [CrossRef]
27. Ravichandran, G.; Subhash, G. Critical appraisal of limiting st rain rates for compression testing of ceramics in a split Hopkinson pressure bar. *J. Am. Ceram. Soc.* **2010**, *77*, 263–267. [CrossRef]
28. Zhou, F.H.; Wang, L.L.; Hu, S.S. On the effect of stress non-uniformness in polymer specimen of SHPB tests. *J. Exp. Mech.* **1992**, *7*, 23–29.

29. Lee, W.-S.; Liu, C.-Y. The effects of temperature and strain rate on the dynamic flow behaviour of different steels. *Mater. Sci. Eng. A* **2006**, *426*, 101–113. [CrossRef]

30. Zhou, T.; Wu, J.; Che, J.; Wang, Y.; Wang, X. Dynamic shear characteristics of titanium alloy Ti-6Al-4V at large strain rates by the split Hopkinson pressure bar test. *Int. J. Impact Eng.* **2017**, *109*, 167–177. [CrossRef]

31. Wu, X.; Li, L.; Liu, W.; Li, S.; Zhang, L.; He, H. Development of adiabatic shearing bands in 7003-T4 aluminum alloy under high strain rate impacting. *Mater. Sci. Eng. A* **2018**, *732*, 91–98. [CrossRef]

Article

Novel Approach in Fracture Characterization of Soft Adhesive Materials Using Spiral Cracking Patterns

Behzad Behnia [1,*] and Matthew Lukaszewski [2]

1   Department of Civil and Environmental Engineering, Clarkson University, Potsdam, NY 13699, USA
2   Department of Electrical and Computer Engineering, Clarkson University, Potsdam, NY 13699, USA; lukaszmp@clarkson.edu
*   Correspondence: bbehnia@clarkson.edu; Tel.: +1-(315)-268-6533

Abstract: A novel approach for the fracture characterization of soft adhesive materials using spiral cracking patterns is presented in this study. This research particularly focuses on hydrocarbon polymeric materials, such as asphalt binders. Ten different asphalt materials with distinct fracture characteristics were investigated. An innovative integrated experimental–computational framework coupling acoustic emissions (AE) approach in conjunction with a machine learning-based Digital Image Analysis (DIA) method was employed to precisely determine the crack geometry and characterize the material fracture behavior. Cylindrical-shaped samples (25 mm in diameter and 20 mm in height) bonded to a rigid substrate were employed as the testing specimens. A cooling rate of $-1\ °C/min$ was applied to produce the spiral cracks. Various image processing techniques and machine learning algorithms such as Convolutional Neural Networks (CNNs) and regression were utilized in the DIA to automatically analyze the spiral patterns. A new parameter, "Spiral Cracking Energy (E_{Spiral})", was introduced to assess the fracture performance of soft adhesives. The compact tension (CT) test was conducted at $-20\ °C$ with a loading rate of 0.2 mm/min to determine the material's fracture energy (G_f). The embrittlement temperature (T_{EMB}) of the material was measured by performing an AE test. This study explored the relationship between the spiral tightness parameter ("b"), E_{Spiral}, G_f, and T_{EMB} of the material. The findings of this study showed a strong positive correlation between the E_{Spiral} and fracture energies of the asphalt materials. Furthermore, the results indicated that both the spiral tightness parameter ("b") and the embrittlement temperature (T_{EMB}) were negatively correlated with the E_{Spiral} and G_f parameters.

Keywords: soft adhesives; asphalt materials; spiral cracking patterns; fracture characterization; acoustic emission; convolutional neural networks (CNNs)

Citation: Behnia, B.; Lukaszewski, M. Novel Approach in Fracture Characterization of Soft Adhesive Materials Using Spiral Cracking Patterns. *Materials* **2023**, *16*, 7412. https://doi.org/10.3390/ma16237412

Academic Editors: Li Ai, Feng Guo, Xinxiang Zhang and Honglei Chang

Received: 4 November 2023
Revised: 22 November 2023
Accepted: 26 November 2023
Published: 29 November 2023

1. Introduction

As a common phenomenon in materials, fractures can create various cracking patterns. It is especially prevalent in thin coatings under residual stresses, forming a network of cracking similar to the patterns observed in dried mud layers and old paintings, see Figure 1a. Some of the well-studied examples of fractures are straight-type cracks, also known as the mud (channeling) cracks, commonly observed in fragmented dried out fields, dried mud layers, coatings, and paintings. In addition to mud fracture patterns, which are the most common type, rare spiral-shaped cracking patterns have been observed in some soft adhesive materials, see Figure 1b.

Spiral cracks usually occur in a thin layer of soft materials coated on rigid substrates. Some distinct processes such as the drying, cooling, syneresis, or stretching of a substrate are attributed as the cause of the formation of this type of crack. Figure 2 schematically illustrates the mechanism of spiral crack development. The deformation mismatch between the coating layer and the substrate induces biaxial tensile stresses within the coating/substrate system, which will act as the driving force behind the crack propagation in the material.

As the magnitude of the induced stresses increases, first a network of channeling cracks (i.e., mud cracks) appears hierarchically, dividing the layer into several polygonal sections. After the formation of mud cracks, the partial detachment of the material from the substrate (partial delamination) takes place within the polygonal sections. A delamination front begins from the edge of the polygonal cell and grows toward the center of the cell. While partial delamination occurs, the spiral crack starts at the border of the adhering region of the fragmentation from some initial imperfection or a weak interface and propagates along a spiral trajectory.

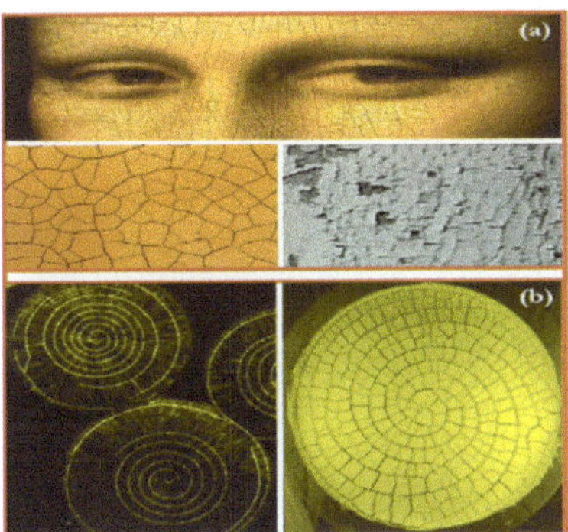

Figure 1. Various fracture patterns in materials. (**a**) Mud cracking patterns. (**b**) Spiral fracture patterns.

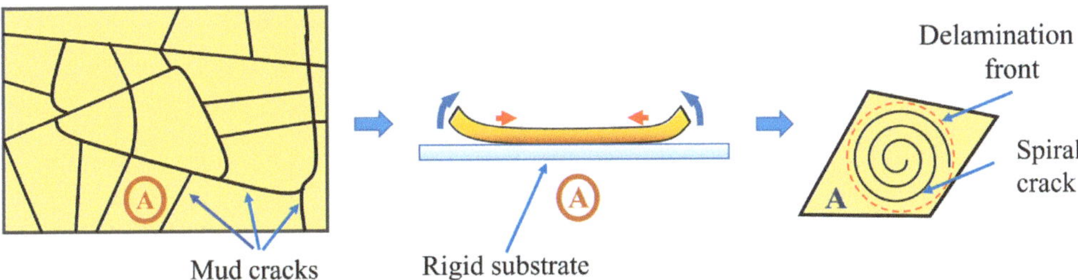

Figure 2. Mechanism of formation of spiral cracks in coating/substrate systems.

Dillard et al. investigated spiral cracks in LaRC™-TPI adhesive (a thermoplastic polyimide material). They found rapid cracking in the thin layer of this material when it was exposed to solvents, such as acetone, toluene, diglyme, and methyl ethyl ketone. They reported on the formation of mud cracking patterns dividing the coating into smaller sections, which is followed by the development of spiral cracks growing inward within each adhesive fragment [1,2]. Macnulty explored the occurrence of various types of cracks in biaxially stressed films of certain polymers containing phenylene, bonded to glass substrates. They reported on the formation of spiral cracks in addition to the regular cracking patterns in the material. They found that both spiral cracks accurately followed the logarithmic curve [3]. Hainsworth et al. investigated thin-film fractures in coating/substrate systems consisting of thin hard coatings deposited on a less stiff, hard substrate during depth-sensing indentation testing (e.g., nanoindentation testing). They conducted experiments on thin TiN and NbN monolayers and also on TiN/NbN and TiN/ZrN composites deposited onto hard steel or stainless steel substrates. They found that some of the cracks had a spiral morphology, which was not sensitive to the indentation loading rate [4]. Neda et al. explored the development of spiral cracks during the desiccation of thin films of precipitates on a substrate. The propagating stress front created by the fragments' foldup was found to be the cause of this type of fracture. They showed that spiral cracks generally occur when the propagating speed of the stress front is proportional to the crack speed [5]. Volinsky et al. studied fractures in thin films caused by residual and/or externally applied stresses. Periodic spiral through-thickness cracks were reported in Mo/Si multilayers subjected to three-point bending in a vacuum at a high temperature [6]. Sendova et al. explored four distinctive cracking patterns including spiral cracks within thin silicate sol-gel films. Cracks geometry was found to be a function of the film thickness, curing time, and temperature. Their study concluded that spiral-shaped crack trajectories happen due to the local warping of the film caused by stress nonuniformity developed at the drying stage of the sol-gel material [7]. In another study, Meyer et al. investigated spiral cracking patterns in the Mo/Si multilayers (60 layers) with the thickness in the range of a nanometer deposited on Si substrates. Spiral cracks were observed in the material during uniaxial bending at high temperatures between 300 and 440 °C in a vacuum. The formation of spiral cracks, which was accompanied by Mo/Si multilayer debonding from the substrate, was linked to the combination of thermal, bending, and residual stresses [8]. Yonezu et al. investigated the initiation and propagation of spiral cracks in a thick layer of diamond-like carbon (DLC) deposited on a steel substrate during spherical indentation. They employed integrated acoustic emission (AE) and corrosion potential fluctuation (CPF) testing methods to evaluate the cracking process. Their experiments showed that spiral cracks can develop only within a narrow range of the maximum indentation force. In case the indentation force is below that range, no fracture happens, and if it is above the range, the ring cracks are formed [9]. Marthelot et al. explored the spiral cracking phenomenon and reported that this type of crack can occur below the standard critical tensile load required for mud cracks. They also demonstrated that spiral cracks choose a robust interaction length scale, which could be 30 times the film thickness [10]. Monev investigated the spiral cracking phenomenon in nickel coatings electrodeposited in an acidic media containing hydrogenation-enhancing additives. The results demonstrated that the type of the additives affects the shape of the cracks as well as the tendency of the nickel layers to form cracks [11]. Wu et al. reported on the formation of perfect Archimedes spiral cracks in a colloid film with several hundred nanometers thickness deposited on a glass substrate [12]. Matsuda et al. studied the occurrence of relatively large-sized spiral cracks (with diameters greater than 0.4 mm) on the surface of melt-grown poly(L-lactic acid) (PLLA) spherulites. The spiral patterns developed in the material due to the thermal shrinkage occurring upon cooling after crystallization. The spiral pitch was found to be positively correlated with the thickness of the spherulite, meaning that the increase in the thickness of the spherulite led to an increase in the spiral pitch [13]. Ma et al. studied spiral cracks in drying suspensions of Escherichia coli (E. coli) with different swimming behaviors. They used 2.5 µL of bacterial suspensions deposited on glass substrates for drying. The spiral cracks were observed in the film of the mutant E. coli with tumbling motions, while the circular cracks were found in the consolidating film of the wild-type E. coli. It was demonstrated that the spiral cracks occur due to film delamination that is caused by the strong bending moment [14], see Figure 3. Behnia et al. reported on the formation of spiral fracture patterns inside a thin layer of hydrocarbon polymeric materials such as asphalt binders and further investigated these patterns [15]. They also explored the potential use of spiral cracking patterns for the fracture characterization of asphalt materials at different oxidative aging levels. They reported that the spiral tightness parameter was found to be sensitive to both the oxidative aging level and the performance grade of the asphalt materials [16]. The logarithmic spiral function was found to be the best fit to mathematically represent 3D spirals, see Equation (1), where "A" is the apparent length scale, "b" is the spiral tightness parameter, D_0 is the initial spiral crack depth, θ is the angle from the x-axis, and θ_f is the final angle corresponding to the outmost point of the spiral in the sample:

$$P(r, \theta, Z) = Ae^{b\theta}\cos(\theta)\,\vec{i} + Ae^{b\theta}\sin(\theta)\,\vec{j} + \frac{\theta}{2\theta_f}D_0\,\vec{k} \tag{1}$$

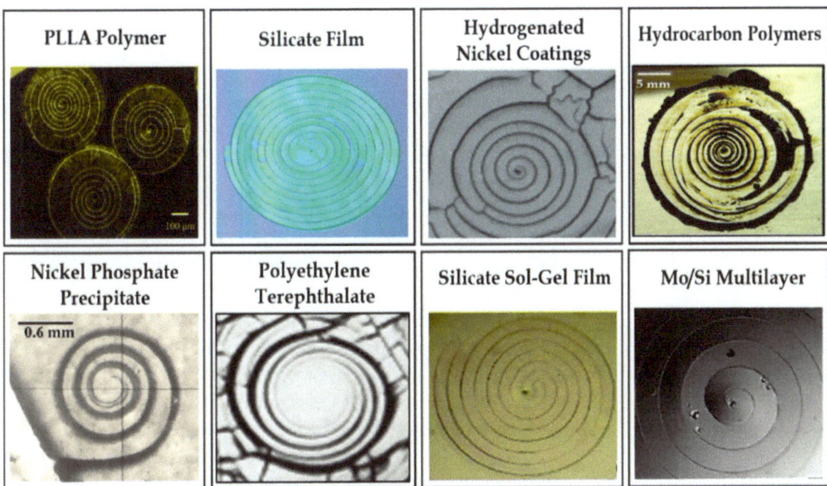

Figure 3. Spiral cracks in different materials [3,5–7,10,11,13,14].

This study presents a transformative and radically different approach using spiral cracking patterns as a powerful diagnostic tool to obtain valuable information about the fracture characteristics of soft adhesives, particularly hydrocarbon polymeric materials. The accurate fracture characterization of soft adhesive materials has remained a challenging task. The implementation of conventional testing methods such as the compact tension (C(T)) test (to evaluate the fracture characteristics) and the Peel test (to measure the adhesive strength) of such materials is quite challenging. Due to the soft nature of these materials, large creep deformations usually occur in the material during the experiment. Moreover, another concerning issue is that the Linear Elastic Fracture Mechanics-based (LEFMs) theories are not suitable to describe the fracture of soft and highly stretchable materials. Research studies have demonstrated that due to large deformations during the fracture process, the stress field near the crack tip in soft materials is significantly different from that used for LEFMs (LEFMs is based on the assumption of infinitesimal deformations) [17].

2. Materials

The present work utilizes hydrocarbon polymeric materials such as asphalt binders to investigate the methodology. It is important to highlight that while the methodology is demonstrated using hydrocarbon polymers, it can be applied to other soft adhesives with minor adjustments in the sample geometry and substrate material selection. Ten different types of asphalt materials (here referred to as *AB1, AB2,…, AB10*) with different fracture properties and the following performance grades (PG) were utilized in this work: *PG58-10, PG58-16, PG58-22, PG58-28, PG58-34, PG64-10, PG64-16, PG64-22, PG64-28,* and the *Styrene-butadiene-styrene (SBS)-modified PG64-22*. The modification of asphalt binders with SBS is a common technique used to enhance the mechanical properties of asphalt materials. The conventional notation for the performance grading of asphalt materials is *PG XX-YY*, where "*XX*" and "*YY*" represent the PG high and low temperatures, respectively. The PG high temperature refers to the average maximum temperature ($^{\circ}$C) that an asphalt road sustains over a seven-day period. On the other hand, the PG low temperature signifies the minimum temperature ($^{\circ}$C) that the asphalt road is likely to encounter throughout its service life [18].

3. Spiral Cracking Experiment

Cylindrical-shaped specimens bonded to a rigid substrate were used as the testing configuration for semi-solid soft adhesives, see Figure 4a. Depending on the type of adhesive material being tested, the diameter (D) and thickness (h) of the specimen as well as the type of substrate material (such as glass, aluminum, etc.) can be different and they should be carefully selected to ensure there are no delamination and no mud cracks within the spiral specimen. For hydrocarbon polymeric materials, various geometries with different thicknesses (ranging from 1 mm to 30 mm) were investigated. The results demonstrated that the cylindrical specimens with D = 25 mm and h = 20 mm bonded to an aluminum substrate showed the most promising results (i.e., no debonding and no mud cracks). The lack of mud cracks in the proposed testing configuration is an advantage that minimizes the variation in the spiral results, thereby enhancing repeatability. Through this investigation, it became evident to the authors that the presence of mud cracks within the sample introduces complexity to the analysis, significantly increasing variations in the results (i.e., reduces repeatability). The fracture-originated acoustic emissions (AE) signals are used for determining the overall 3D geometry of a spiral crack as well as finding the characteristic parameters of the spirals, such as the spiral tightness parameter. In the case of using other sample geometries such as a prismatic shape, the formation of spiral cracks is accompanied by the occurrence of mud cracks in the specimen. The length, formation location, and orientation of mud cracks exhibit variability from one sample to another, introducing complexity to the analysis of the AE signals. The recorded AE signals in the samples with mud cracks can be categorized into two parts: One part is the AE signals originating from the spiral crack (this part is used in the analysis). Another part is the AE signals stemming from the mud cracks. Variations in the geometric characteristics of the mud cracks result in notable differences in the number of AE signals, their energy content, amplitude, frequency content, etc. Consequently, this variability contributes to an increase in the Coefficient of Variation (CoV%) of the results and makes the interpretation of the spiral cracking results complicated.

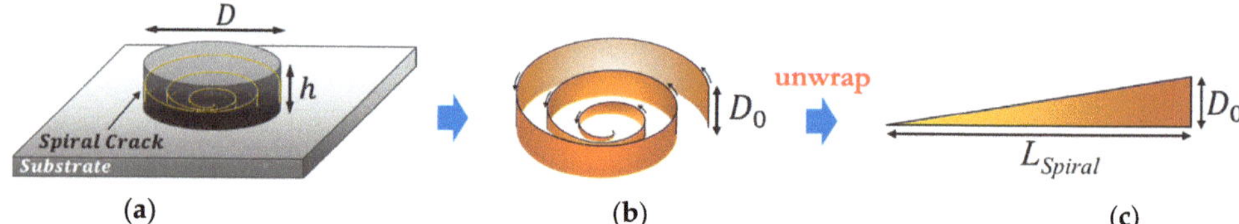

Figure 4. (**a**) Testing configurations for semi-solid soft adhesives. (**b**) Three-dimensional spiral crack developed in the sample. (**c**) Unwrapped spiral crack in the form of a triangle.

Spiral cracks are produced in the lab under a controlled condition. Depending on the type of material, they could be thermally induced, solvent-induced, or caused by drying. In the case of hydrocarbon polymeric materials, spiral cracks are thermally induced and produced by cooling the sample. In the preliminary study, various cooling rates including 0.3, 0.5, 1, 1.5, and 2 °C/min were investigated. The results showed that cooling rates higher than 1 °C/min resulted in sample delamination from the substrate before the spiral crack even gets a chance to develop. Thus, the average cooling rate of 1 °C/min was applied for testing the asphalt materials. The asphalt samples were prepared by pouring heated material into a cylindrical-shaped silicon mold mounted on the substrate. The samples were allowed to reach room temperature before the testing. To conduct the experiment, the sample was cooled down from 0 to −50 °C at the average rate of 1 °C/min. We selected −50 °C to make sure the spiral crack was fully developed inside the sample. The results showed that the formation of a spiral crack in the asphalt samples subjected to a 1 °C/min cooling rate was complete at temperatures ranging from −40 °C to −50 °C for various types of asphalt. As the temperature reduces, differential thermal contraction between the rigid substrate and asphalt creates thermally induced stresses within the material, which eventually leads to the formation of an inward-growing spiral crack nucleating from some initial imperfection or a weak interface near the edge of the specimen.

The results showed a gradual reduction in the spiral crack depth (penetration depth) from almost 50% of the sample thickness (h) at the edge to almost zero at the center of the specimen, see Figure 4b. This phenomenon could be linked to a gradual reduction in the amount of stored strain energy as the crack spirals from the edge toward the center of the sample. During the fracture process, the stored strain energy in the specimen is consumed for the creation of new fractured surfaces. At the beginning of the fracture process, the stored strain energy in the specimen is at the highest level, leading to the nucleation of the spiral crack at the interface, with a maximum penetration depth, D_0. As the spiral crack grows, the strain energy is gradually diminished to create new fractured surfaces. As a result, the crack penetration depth continuously decreases until it reaches almost zero at the center of the sample. The careful inspection of a fully grown 3D spiral showed that the gradual change in the spiral crack depth is almost linear. Figure 4c schematically shows the hypothetical unwrapped shape of a 3D spiral crack in the form of a triangle.

4. Integrated Experimental and Computational Framework

An overview of the integrated experimental and computational framework consisting of the multi-sensor *AE* method coupled with the Digital Image Analysis (*DIA*) approach to evaluate the fracture characteristics of soft adhesive materials is illustrated in Figure 5 (the technical details of these methods are provided in Sections 4.1 and 4.2). In this approach, the total area of the spiral-shaped fractured surfaces ($A_{Fracture}$) inside the specimen is calculated through the application of the coupled *AE-DIA* approach, where the total length of the spiral crack (L_{Spiral}) is measured using *DIA* and the initial depth of the spiral crack D_0 is determined using the *AE* source location. For the *DIA* analysis, a novel machine learning-based image processing framework is applied. The AE energies of individual events (i.e., the *AE* event is a rapid physical change, such as microcracks appearing as an acoustic signal) will be added up to measure the total amount of released *AE* energy due to the formation of a spiral crack within the sample. A new parameter (index) called the spiral cracking energy (E_{Spiral}) is introduced, which is defined as the amount of released *AE* energy per unit of the newly formed fracture surface area of the spiral cracks. It should be noted that this fracture index is not the same as the fracture energy of the material due to the fact that the measured *AE* energy is not equal to the strain energy released during crack propagation. During the spiral cracking process, part of the strain energy released in the specimen is used to create new fractured surfaces, and the rest is released as transient elastic waves, which can be picked up by *AE* sensors (a portion of transient waves can be dissipated by attenuation before reaching *AE* sensors). The former is related to the fracture energy and the latter is captured by the *AE* method. The E_{Spiral} index quantifies the fracture resistance of soft adhesive materials using their *AE* activities during the spiral cracking. The unit of E_{Spiral} is $V^2.\mu s/mm^2$ and it can be calculated by dividing the total released *AE* energies of the fracture-induced signals by the total fractured surface area within the sample, Equation (2), where N is the total number of recorded *AE* signals, $V_i(t)$ is the voltage of the *i*th recorded signal in volts, and $A_{Fractured}$ is the total surface area of the fractured faces measured from the *AE-DIA* analysis.

$$E_{spiral} = \frac{\sum_{i=1}^{N} \int_0^t V_i^2(t)dt}{A_{Fractured}} \tag{2}$$

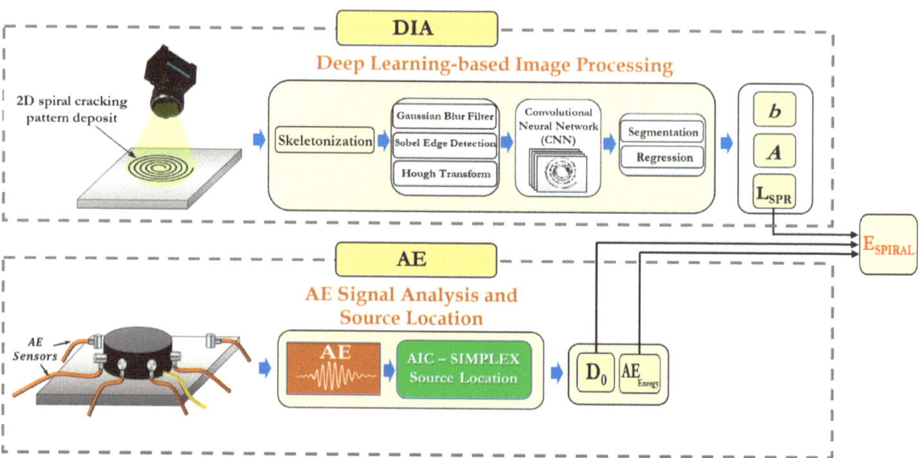

Figure 5. Integrated *EA-DIA* approach to assess fracture characteristics of soft adhesives.

4.1. Multi-Sensor Acoustic Emission Testing

The multi-sensor acoustic emission (*AE*) technique was employed to continuously monitor the acoustic activities of the specimen during the course of the spiral cracking experiment and also to measure the initial depth of the spiral crack, D_0, see Figure 6. To prepare the cylindrical asphalt samples with a diameter of 25 mm and height of 20 mm, hot liquified asphalt at 135 °C was poured into a cylindrical silicon rubber mold mounted securely on an aluminum plate. To facilitate the sample demolding process, the silicon model was covered with Teflon tape. To ensure proper bonding between the asphalt and the substrate, the aluminum plate was heated to 135 °C before pouring the binder. The sample was let to cool down to room temperature and then placed in the freezer (set at 0 °C) for a few minutes to make the demolding process easy. A spiral crack developed inside the material by cooling the sample from room temperature to −50 °C at an average cooling rate of −1 °C/min.

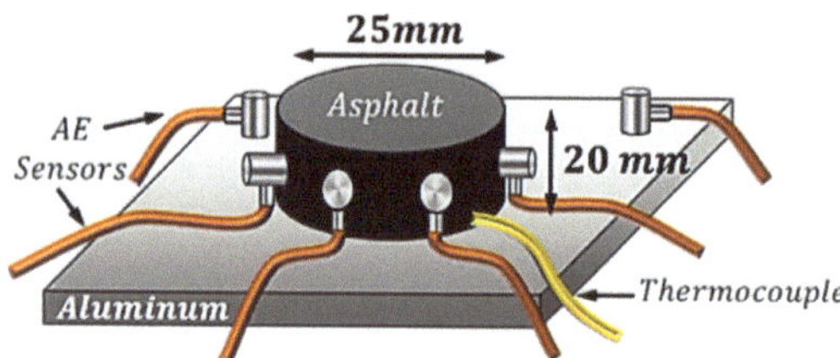

Figure 6. Testing specimens used for hydrocarbon polymeric materials.

AE sensors with a relatively flat response over the target frequency range capable of working properly in the target range of the test temperatures were utilized. For the asphalt materials, broadband *AE* sensors with flat responses in the frequency ranging 20 kHZ–1 MHz were used. To minimize the extraneous noise (i.e., separating genuine fracture-originated *AE* signals from noise), the signals were pre-amplified to 20 dB using broadband pre-amplifiers. The signals were then further amplified by 21 dB (for a total of 41 dB) and filtered using low-pass (LPF) and high-pass (HPF) filters of 500 kHz and 20 kHz, respectively, with the Fracture Wave Detector (FWD) signal condition unit. The signals were digitized using a 16-bit analog-to-digital converter and a sampling frequency of 1 MHz and a length of 2048 points per channel per acquisition trigger. At the postprocessing stage, all the AE signals with energy lower than 4 V^2-μs were filtered out. Moreover, the standard pencil-lead break (PLB) test was performed routinely before conducting the experiments in order to calibrate the *AE* system as well as the *AE* sensors and to make sure the variation within the *AE* channels was negligible.

As the spiral crack propagates inward, new fractured surfaces are formed, which are accompanied by the release of stored strain energy in the form of transient mechanical stress waves inside the specimen. The *AE* piezoelectric sensors mounted on the surface of the specimen as well as on the substrate adjacent to the specimen will continuously monitor and detect these mechanical waves and convert them into *AE* signals. The *AE* signals carry valuable information about the spiral fracture process occurring in the material. Therefore, the *AE* signals were recorded, carefully analyzed, and the key features of the signals such as the signal amplitude, energy, frequency, hit counts, arrival time, and duration were extracted.

In addition, the *AE* source location technique was applied to measure the starting penetration depth of the spiral crack (D_0). An integrated approach combining the two-step *Akaike Information Criterion (AIC)* method [19] and Simplex algorithm (also known as the Nelder–Mead algorithm) was implemented. The *AIC* approach was used for the precise automatic determination of the time of arrival (*ToA*) of the *AE* signals. The autoregression-based *AIC* function divides the *AE* signal, $\{X_1, X_2, \ldots, X_N\}$, into two vectors at the time k: $\{X_1, X_2, \ldots, X_k\}$ and $\{X_{k+1}, X_{k+2}, \ldots, X_N\}$. The method then compares the signal variance of the prediction errors before and after the time k in a predetermined time window. The *AIC* value at point i $=$ k is calculated using Equation (3), where N is the number of amplitudes of a digitized *AE* wave; X_i is an amplitude of a signal ($i = 1, 2, .., N$); $\mathrm{var}(X[1, k])$ is the variance of X between X_1 and X_k; and $\mathrm{var}(X[k+1, N])$ is the variance of X between X_{k+1} and X_N. The *ToA* of an *AE* signal is the time at which the *AIC* function becomes a global minimum. In the two-step *AIC* process, in the first step, the global minimum of the *AIC* function is employed to obtain the first estimation of the *ToA*. In the second step, the time interval is narrowed down and focused on the neighborhood of the first *ToA* estimation. The final value of the *ToA* of the signal is computed using the global minimum of the recalculated *AIC* function.

$$\mathrm{AIC}_k = k.\log\{\mathrm{var}(X[1, k])\} + (N - k - 1).\log\{\mathrm{var}(X[k+1, N])\} \tag{3}$$

After calculating the *ToA*s of the *AE* signals, the Simplex algorithm was used for the source location. In this method, for any point in the medium, an error (E) was computed by comparing the calculated and observed arrival times. As such, an error space is created, in which each point is an error associated with a point in the specimen. The point with the minimum error is the event location. Equation (4) is used to calculate the location error using the least squares method (L_2 *norm*), where E is the least squares error (L_2 *norm error*); t_i is the *ToA* obtained from the *AIC* picker; tt_i is the travel time from the location of interest to the *i*th sensor; n is number of *AE* sensors; m is number of equations; and q is the degree of freedom. The error space is created by computing E for every point in the medium and the point with the minimal error is localized as the source location of the *AE* event.

$$E = \sqrt{\frac{\sum\left[\left(t_i - \frac{\sum t_i}{n}\right) - \left(tt_i - \frac{\sum tt_i}{n}\right)\right]^2}{m - q}} \tag{4}$$

The accuracy of the source localization was assessed using the *Euclidean Distance Error (EDE)*, in which the *PLB* procedure (*Hsu–Nielsen* source) is used to artificially generate *AE* signals at various locations of the sample. The source of the *AE* activities was located and compared with the actual locations. The *EDE* was measured by calculating the distance between the actual source location and the estimated source location. The results showed around 92% accuracy of the source location approach in the spiral specimens.

In addition to determining the crack's initial penetration depth, the *AE* source location was also implemented to measure the 3D shape and size of the fracture process zone (*FPZ*) ahead of the spiral crack tip. The results showed that unfortunately the source location did not have a sufficient spatial resolution to map out the accurate 3D geometry of the *FPZ* in the spiral specimens. Figure 7 illustrates a typical map of *AE* activities within the profile of the spiral specimen obtained through an *AE* source location analysis. The gray dashed lines represent the boundaries of the spiral specimen (25 mm in diameter and 20 mm in height), while the red dots indicate the locations of the *AE* events that occurred within the sample. The results of the *AE* source location analysis demonstrate that the initial depth of the spiral crack (D_0) can be accurately measured, see Figure 7. However, due to the limitations in the spatial resolution of the *AE* source location, it was not possible to precisely detect the exact location of the crack front and measure its depth as it spirals toward the sample center.

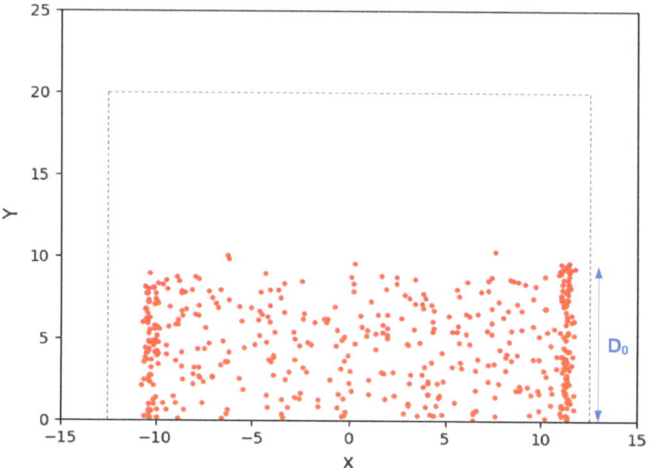

Figure 7. Map of AE events activities (red dots) within spiral cracking sample.

The *AE* test was also used to measure the embrittlement temperature (T_{EMB}) of the material. The T_{EMB} is considered as the starting point of thermally induced damage in the material. By analyzing the *AE* signals in conjunction with the recorded test temperatures, the temperature corresponding to the occurrence of the *AE* event with the first peak energy level was identified. This temperature is used as the T_{EMB} of the material.

4.2. Machine Learning-Based Framework for Digital Image Analysis (DIA)

The pipeline of the framework used for the automated *DIA* of the spiral cracking patterns is illustrated in Figure 8. It consists of various image processing techniques along with the *Convolutional Neural Networks* (*CNNs*) machine learning approach. In the first step, image skeletonization was applied, in which a spiral cracking image was reduced to a one-pixel-thick skeleton of a crack path and the output image was converted into grayscale. Skeletonization was performed to accelerate the analysis by providing a light skeleton for image processing instead of an otherwise computationally expensive analysis on the original image. After skeletonization, the Gaussian blurring filter and Sobel method were applied to remove the inhomogeneous image background illuminations as well as for the edge detection of the cracking patterns, respectively. This was followed by the application of the Hough transform method to detect the spiral shapes in the image. This approach is capable of detecting shapes in images even if those shapes are slightly broken or distorted [20].

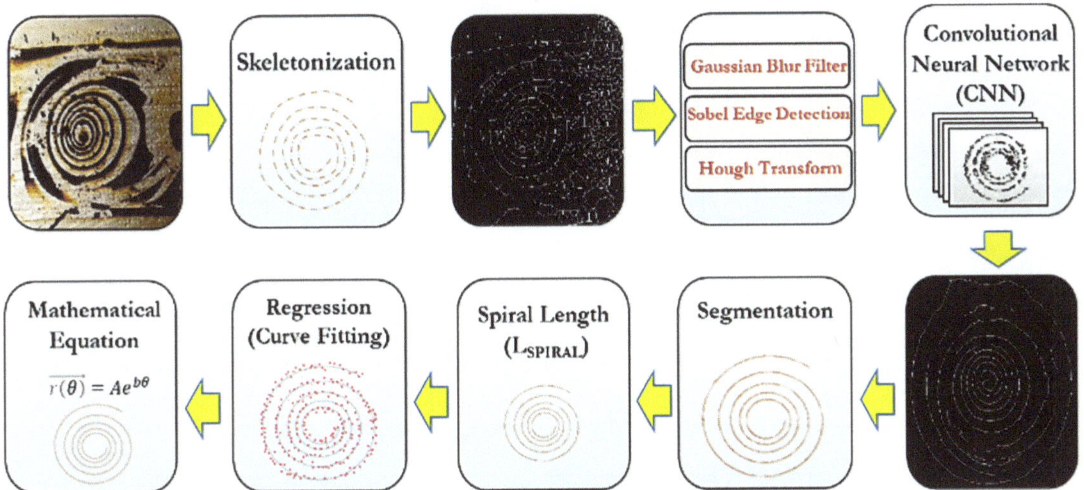

Figure 8. Pipeline of the ML-based DIA framework.

The *CNN* algorithm was implemented for noise reduction in the skeletonized images. This algorithm is ideal for image processing tasks as it significantly reduces the number of required weights for neurons in the model by using tiling regions, each with the same shared weights. The *CNN* autoencoder compresses spiral cracking images in a series of convolutions and then reverses the process during the decoding step. The hidden layers of the *CNN* perform convolutions generating a feature map, which contributes to the input of the next layer. Each convolutional layer is followed by a Rectified Linear Unit (*ReLU*) layer. A regular feedforward neural network consisting of a couple of fully connected layers (+*ReLUs*) is added at the beginning of the stack and the final layer outputs the prediction using the *Softmax* activation function. To properly train the *CNN* model in addition to real spiral crack images, a large number of synthetic images of spiral patterns with various shapes and sizes were generated. Random artificial noise was added to both the real and synthesized spiral images. A combination of real and synthesized images with and without artificial noise was used for training the *CNN*. Once the noise was removed from the images, the segmentation technique was performed to separate and extract the spiral crack path from the original and was followed by measuring the total length of the spiral crack (L_{Sprial}). Finally, a regression analysis was employed to determine the spiral parameters: "*A*" and "*b*".

5. Results and Discussion

Table 1 summarizes the spiral cracking parameters from the *AE-DIA* approach, the fracture energy (G_f) values obtained from compact tension ($C(T)$) test, and the embrittlement temperature (T_{EMB}) from the AE test. Each data point in Table 1 is the average of six to eight testing replicates. Performing a $C(T)$ test for semi-solid adhesives is a challenging task due to the soft nature of these materials. In this study, various temperatures ranging from 0 °C to −30 °C and the loading rates (0.1-1 mm/min) were carefully explored. It was observed that for the asphalt binders, the $C(T)$ tests performed at −20 °C with a loading rate of 0.2 mm/min were satisfactory. The $C(T)$ test was performed in accordance with ASTM E399-05 to determine the fracture energy of the asphalt binders [21]. To prepare the $C(T)$ samples, hot liquified asphalt at 135 °C was poured into a silicon rubber mold and the samples were left to cool down. After reaching room temperature, the samples were placed in the freezer for 5 min to make the demolding process easy. The $C(T)$ samples were conditioned for two hours at −20 °C before conducting the experiment. The fracture energy (G_f) of the material was computed by dividing the calculated area under the load–CMOD curve by the fractured surface area. Some fracture energy data are missing in Table 1 primarily because the authors encountered sample failure during the *C(T)* tests, preventing them from successfully conducting the experiments for those materials.

Table 1. Experimental results for ten different asphalt materials.

Material ID	Performance Grade (PG)	Φ (deg)	b	CoV (%)	E_{Spiral} ($v^2 \cdot \mu s/mm^2$)	CoV (%)	AET_{EMB} (°C)	CoV (%)	G_f (J/m^2)	CoV (%)
AB-1	PG 58-10	3.907	0.0683	9.2	147	11.5	−15.2	12.6	-	-
AB-2	PG 58-16	2.674	0.0467	11.6	435	15.9	−29.6	6.5	2.42	18.3
AB-3	PG 58-22	2.742	0.0479	7.3	411	8.5	−31.9	9.1	8.27	13.9
AB-4	PG 58-28	2.411	0.0421	13.8	761	10.1	−34.3	13.3	13.09	16.2
AB-5	PG 58-34	1.919	0.0335	10.5	1092	9.7	−39.8	10.5	34.11	21.7
AB-6	PG 64-10	3.388	0.0592	5.7	185	14.9	−17.4	8.6	-	-
AB-7	PG 64-16	2.594	0.0453	9.3	335	10.2	−28.0	12.1	-	-
AB-8	PG 64-22	2.525	0.0441	12.2	455	8.4	−31.1	7.2	3.14	15.5
AB-9	PG 64-28	2.382	0.0416	11.4	447	13.5	−30.6	15.7	19.32	22.1
AB-10	PG 64-22+SBS	2.050	0.0358	16.5	841	18.6	−34.5	10.4	37.17	16.5

The results showed that the E_{Spiral} values were positively correlated with the fracture energies of the asphalt materials, meaning that the average amount of released *AE* energy per unit area of spiral cracks is higher for materials with higher resistance against cracking. A comparison of the E_{Spiral} and G_f with the spiral tightness, "*b*", values demonstrated a different trend in which both E_{Spiral} and G_f were negatively correlated with "*b*". This could be explained through understanding the mechanism behind the formation and propagation of spiral cracks. It is hypothesized that a spiral crack front selects its trajectory in a direction where it could maximize the stored strain energy release rate [22]. As such, spiral cracks have a constant pitch angle, meaning that the orientation of the crack front is constantly bending away from the instantaneous propagation direction. The spiral pitch angle is lower in high fracture-resistant materials, because it is more difficult for the crack front to bend away from its instantaneous propagation direction in high fracture-resistant materials. A lower spiral pitch angle (φ) results in a lower spiral tightness parameter "*b*" ($b = \tan(\varphi)$). The findings of this study showed the dependency of the G_f of asphalt to its PG temperatures, particularly to the PG low temperature (*PGLT*) of the material, where binders with a lower *PGLT* exhibited higher fracture energies [15,16]. Additionally, it was observed that the *PGLT* of the binder has a significant effect on its "*b*" value. Generally, the lower the *PGLT* of a binder, the higher its low-temperature cracking resistance (higher fracture energy). As a result, asphalt materials with a lower *PGLT* tend to have lower "*b*" values. Furthermore, it is noteworthy that the asphalt modified with *SBS* demonstrates the highest fracture energy and a relatively low spiral tightness parameter. This outcome can be attributed to the influence of the *SBS* polymer modification to enhance the resistance of the material against cracking.

The *AE* results suggest a strong correlation between the *AE* embrittlement temperature (T_{EMB}), spiral tightness parameter, and fracture energy of the asphalt materials. The observations reveal a positive correlation between the T_{EMB} of the binders and the "*b*" value, while a negative correlation exists between the T_{EMB} and the fracture energy of the material. A summary of the observed correlations between the spiral tightness (b), spiral cracking energy (E_{Spiral}), embrittlement temperature (T_{EMB}), and fracture energy (G_f) for the hydrocarbon polymeric materials is presented in Table 2.

Table 2. Statistical correlations between spiral tightness (b), spiral cracking energy (E_{Spiral}), embrittlement temperature (T_{EMB}), and fracture energy (G_f) for hydrocarbon polymeric materials.

		Correlation			
		b	E_{Spiral}	T_{EMB}	G_f
Correlation	b	+	-	+	-
	E_{Spiral}	-	+	-	+
	T_{EMB}	+	-	+	-
	G_f	-	+	-	+

Typical plots of the *AE* results including the cumulative *AE* hit counts and *AE* energy versus temperature are presented in Figure 9. The *AE* response of the material during spiral cracking showed four distinct regions: region#1 (pre-cracking region), region#2 (transition region from quasi-brittle to brittle state), region#3 (stable cracking region), and region# 4 (fully cracked region). The magnitude of the emitted *AE* energies is highest when the spiral crack starts to propagate (at the start of region#2). As the crack continues to grow, the magnitude of the emitted *AE* energies gradually tapers off until it reaches almost zero at the end of region#4 when the spiral crack is fully developed. This can be linked to a gradual reduction in the stored strain energy (as the driving force behind crack propagation) when the crack front advances from the edge toward the center of the sample. Figure 4 schematically shows the 3D spiral cracking pattern with the gradual reduction in crack penetration depths. During the fracture process, the stored strain energy in the specimen is consumed for the creation of new fractured surfaces. At the end of region#1 (the pre-cracking region), the stored strain energy in the specimen is at the highest level, leading to the nucleation of the spiral crack at the interface with a maximum penetration depth, D_0. As the crack grows, the strain energy is gradually diminished to create new fractured surfaces. As a result, the crack penetration depth continuously decreases until it reaches almost zero at the center of the sample. A visual inspection of the fully grown 3D spiral showed that the gradual change in the spiral crack depth is almost linear. Figure 4 demonstrates the hypothetical unwrapped shape of the 3D spiral crack in the form of a triangle.

A further analysis of the *AE* signals was performed to investigate other important *AE* parameters, such as the signal duration time, signal rise time (*RT*), frequency content, rise angle (*RA = Signal rise time/Signal peak amplitude*), and average frequency (*AF = AE hit counts/Signal duration time*). An interesting observation was that the *AE* signals recorded at the beginning of the spiral crack formation had a low rise time, high amplitude values, and a high frequency (low *RA* and high *AF*) (this type of signal is usually observed in fracture Mode I), while the signals recorded at lower temperatures exhibited a longer duration time, longer rise times, a low amplitude, and a lower frequency (high *RA* and low *AF*) (these types of signals are typically observed in fracture *Mode II*), see Figures 9c and 10. These findings suggest that during the course of the spiral cracking process, the fracture mode changes from *Mode I* to *Mode II*. It was observed that the change in the *AF* and *RA* values mostly occurred at temperatures near the glass transition temperature (T_g) of the material, where the material behavior gradually changes from quasi-brittle to a brittle state (the T_g values for the asphalt materials utilized in the present study were in the range of $-25\ °C$ to $-30\ °C$). As a result, it is hypothesized that a change in the fracture mode of spiral cracks happens at temperatures close to the glass transition temperature of the material (within the transition region (region#2)). A more in-depth investigation is required to further explore this hypothesis.

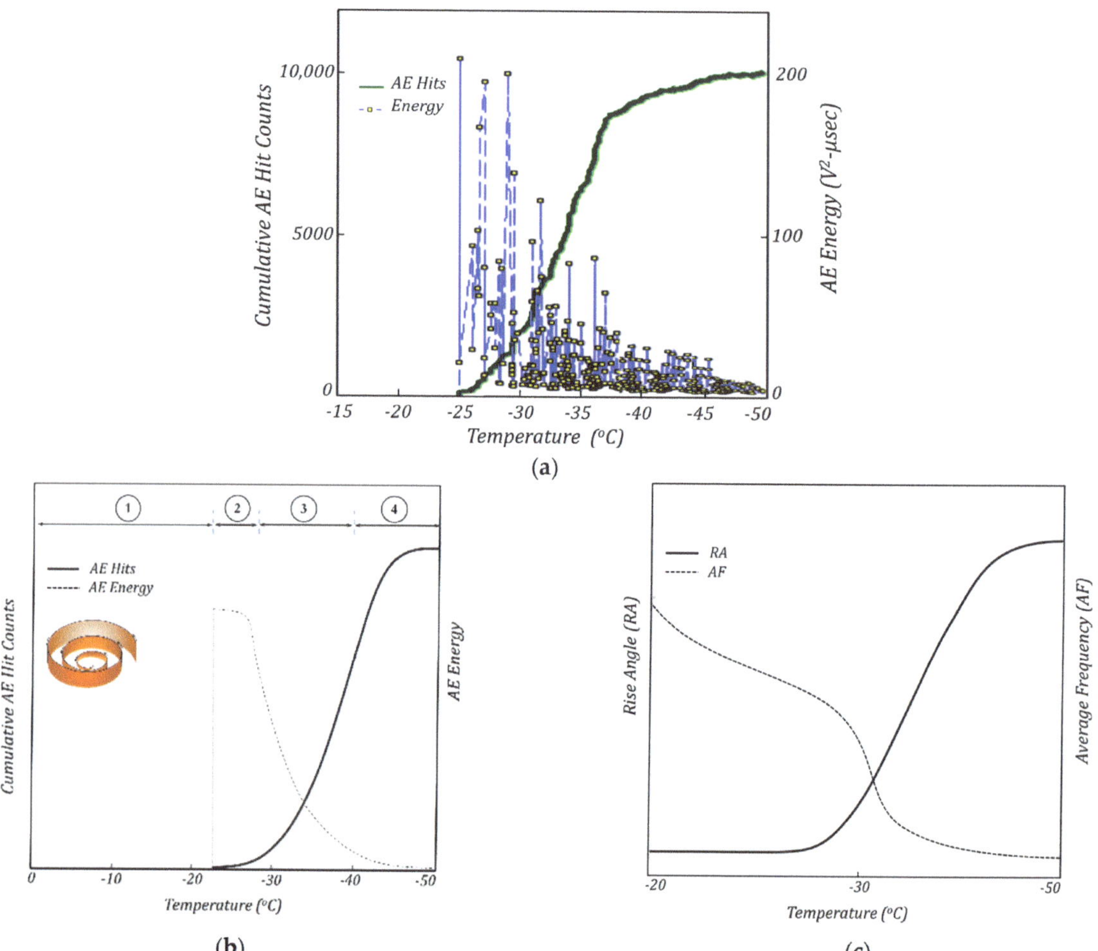

Figure 9. (**a**) Typical plot of cumulative AE hit counts and AE events energy versus temperature. (**b**) Typical envelope locus of AE hit counts and AE energies vs. temperature. (**c**) AE rise angle and average frequency vs. temperature.

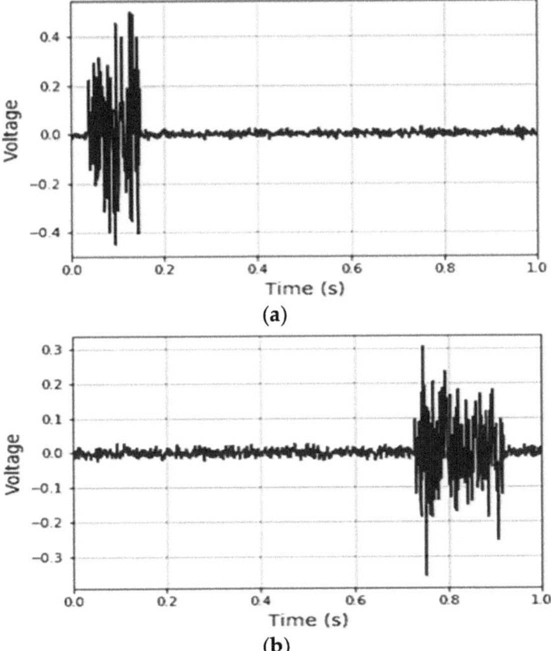

Figure 10. (**a**) Typical *AE* signal at temperatures above the glass transition temperature of the material: low rise time, high amplitude values, and high frequency (low *RA* and high *AF*). (**b**) Typical *AE* signal at temperatures below the glass transition temperature of the material: longer waveforms, with longer rise times, low amplitude, and lower frequency (high *RA* and low *AF*).

6. Conclusions

The present work investigates a novel approach for the fracture characterization of soft adhesives using spiral cracking patterns. Ten different hydrocarbon polymeric materials with various fracture characteristics were utilized in this study. Cylindrical-shaped samples (with a 25 mm diameter and 20 mm height) bonded to a rigid substrate (aluminum plate) were used as the testing specimens for the hydrocarbon polymeric materials. A spiral crack formed inside the specimen by cooling it from room temperature to -50 °C at an average rate of -1 °C/min. An integrated experimental–computational framework coupling the multi-sensor *AE* and *DIA* approaches was employed to determine the spiral cracking parameters, such as the spiral tightness parameter. An efficient image processing framework using various image processing and machine learning techniques such as *CNNs*, skeletonization, segmentation, and regression were used in the *DIA* for the automatic analysis of the spiral pattern images. A new parameter called the *"Spiral Cracking Energy (E_{Spiral})"* was introduced to evaluate the fracture performance of the soft adhesives.

In addition to the spiral cracking parameters, some other critical parameters used in the low-temperature cracking evaluation of the hydrocarbon polymeric materials such as the fracture energy and the embrittlement temperature were measured. The compact tension (CT) test was conducted at -20 °C with a loading rate of 0.2 mm/min to determine the fracture energy (G_f) of the material. The embrittlement temperature (T_{EMB}) of the material was determined by performing an AE test. A summary of the observed correlations between the spiral tightness (b), spiral cracking energy (E_{Spiral}), embrittlement temperature (T_{EMB}), and fracture energy (G_f) are presented in the following:

- Exploring the relationship between the E_{Spiral} and G_f values of the hydrocarbon polymeric materials showed a strong positive correlation between these two parameters, where an increase in one leads to an increase in the other one, and vice versa.
- Additionally, the results indicated that the spiral tightness parameter ("b") was negatively correlated with both the E_{Spiral} and G_f parameters for the hydrocarbon polymeric materials, meaning that the increase in the spiral tightness parameter is associated with a decrease in the E_{Spiral} and G_f parameters, and vice versa.
- Investigating the relationship between the T_{EMB} and spiral tightness parameter revealed a positive correlation between the T_{EMB} and the "b" value in the hydrocarbon polymeric materials.
- The experimental results also showed a negative correlation between the E_{Spiral} and G_f parameters and the T_{EMB} of the hydrocarbon polymeric materials.

The characteristic parameters of the AE signals such as the duration time, rise time (RT), frequency content, rise angle (RA), and average frequency (AF) were analyzed. It was observed that at the beginning of the spiral cracking process, the signals exhibited a low RA and high AF (usually observed in fracture Mode I). On the other hand, later, at lower temperatures (lower than the glass transition temperature), the recorded AE signals were mostly of a high RA and low AF (typically observed in fracture Mode II). It is hypothesized that a change in the fracture mode may happen at temperatures near the glass transition temperature (T_g) of the material, at which the material behavior changes from quasi-brittle to brittle.

Given the current challenges in carrying out conventional fracture tests for soft adhesives, the findings of this limited study suggest that the use of spiral cracking patterns could be considered as a viable alternative for assessing the fracture characteristics of such materials and laying the groundwork for future advancements in this field. Particularly in the field of hydrocarbon polymeric materials, numerous studies have been directed toward the low-temperature cracking characterization of asphalt materials [23–29]. It is recommended that for future studies, the spiral cracking results are compared against those from the existing testing methods, such as the Asphalt Binder Cracking Device (ABCD), Bending Beam Rheometer (BBR), etc. Such a comparative analysis aims to further explore the correlations and establish relationships between different testing approaches, paving the way for enhanced insights into the fracture performance of asphalt materials.

Author Contributions: Conceptualization, B.B.; Methodology, B.B. and M.L.; Software, M.L.; Validation, M.L.; Formal analysis, M.L.; Investigation, M.L.; Data curation, M.L.; Writing—original draft, B.B.; Supervision, B.B.; Project administration, B.B.; Funding acquisition, B.B. All authors have read and agreed to the published version of the manuscript.

Funding: This research study was supported by the National Science Foundation under grant number: 1911732.

Institutional Review Board Statement: Not applicable.

Informed Consent Statement: Not applicable.

Data Availability Statement: Some or all of the data that support the findings of this study are available from the corresponding author upon reasonable request.

Acknowledgments: The authors would like to thank the National Science Foundation and especially its director of the Mechanics of Materials and Structures program, Siddiq Qidwai, for their support.

Conflicts of Interest: The authors declare no conflict of interest.

References

1. Argon, A.S. Distribution of cracks on glass surfaces. *Proc. R. Soc. Lond. A* **1959**, *250*, 482–492.
2. Dillard, D.A.; Hinkley, J.A.; Johnson, W.S.; Clair, T.L.S. Spiral Tunneling Cracks Induced by Environmental Stress Cracking in LaRC™-TPI Adhesives. *J. Adhes.* **1994**, *44*, 51–67. [CrossRef]

3. Macnulty, B.J. Spiral and other crazing and cracking in polymers with phenylene groups in the main chain. *J. Mater. Sci.* **1971**, *6*, 1070–1075. [CrossRef]
4. Hainsworth, S.V.; McGurk, M.R.; Page, T.F. The effect of coating cracking on the indentation response of thin hard-coated systems. *Surf. Coat. Technol.* **1998**, *102*, 97–107. [CrossRef]
5. Néda, Z.; Leung, K.-T.; Józsa, L.; Ravasz, M. Spiral Cracks in Drying Precipitates. *Phys. Rev. Lett.* **2002**, *88*, 095502. [CrossRef]
6. Volinsky, A.A.; Moody, N.R.; Meyer, D.C. Stress-induced periodic fracture patterns in thin films. In Proceedings of the 11th International Congress on Fracture Proceedings, Turin, Italy, 20–25 March 2005.
7. Sendova, M.; Willis, K. Spiral and curved periodic crack patterns in sol-gel films. *Appl. Phys. A* **2003**, *76*, 957–959. [CrossRef]
8. Meyer, D.C.; Leisegang, T.; Levin, A.A.; Paufler, P.; Volinsky, A.A. Tensile crack patterns in Mo/Si multilayers on Si substrates under high-temperature bending. *Appl. Phys. A* **2004**, *78*, 303–305. [CrossRef]
9. Yonezu, A.; Liu, L.; Chen, X. Analysis on spiral crack in thick diamond-like carbon film subjected to spherical contact loading. *Mater. Sci. Eng. A* **2008**, *496*, 67–76. [CrossRef]
10. Marthelot, J.; Roman, B.; Bico, J.; Teisseire, J.; Dalmas, D.; Melo, F. Self-Replicating Cracks: A Collaborative Fracture Mode in Thin Films. *Phys. Rev. Lett.* **2014**, *113*, 085502. [CrossRef]
11. Monev, M. Hydrogenation and cracking of nickel coatings electrodeposited in the presence of brighteners. *Bulg. Chem. Commun.* **2016**, *48*, 73–77.
12. Wu, D.; Yin, Y.J.; Xie, H.M.; Dai, F.L. Archimedes spiral cracks developed in a nanofilm/substrate system. *Chin. Phys. Lett.* **2013**, *30*, 036801. [CrossRef]
13. Matsuda, F.; Sobajima, T.; Irie, S.; Sasaki, T. Spiral crack patterns observed for melt-grown spherulites of poly (L-lactic acid) upon quenching. *Eur. Phys. J. E* **2016**, *39*, 41. [CrossRef] [PubMed]
14. Ma, X.; Liu, Z.; Lin, T.; Zeng, W.; Tian, X.; Cheng, X. Circular and spiral-like crack patterns of desiccated bacterial deposits. In *APS March Meeting Abstracts*; American Physical Society: College Park, MD, USA, 2022; Volume 2022, p. Z03.013.
15. Behnia, B.; Buttlar, W.G.; Reis, H. Spiral cracking pattern in asphalt materials. *Mater. Des.* **2017**, *116*, 609–615. [CrossRef]
16. Behnia, B.; Lopa, R.S. Use of spiral cracking patterns in fracture characterization of asphalt materials. *Constr. Build. Mater.* **2022**, *361*, 129581. [CrossRef]
17. Lazarus, V.; Pauchard, L. From craquelures to spiral crack patterns: Influence of layer thickness on the crack patterns induced by desiccation. *Soft Matter* **2011**, *7*, 2552–2559. [CrossRef]
18. Asphalt Institute. *Superpave Mix Design*; Superpave Series No. 2; Asphalt Institute: Lexington, KY, USA, 2001.
19. Akaike, H. A new look at the statistical model identification. *IEEE Trans. Autom. Control.* **1974**, *19*, 716–723. [CrossRef]
20. Aghdam, H.H.; Heravi, E.J. *Guide to Convolutional Neural Networks*; Springer: New York, NY, USA, 2017; Volume 10, p. 51.
21. *ASTM E399-05*; Standard Test Method for Linear-Elastic Plane-Strain Fracture Toughness KIc of Metallic Materials. ASTM: West Conshohocken, PA, USA, 2012.
22. Xia, Z.C.; Hutchinson, J.W. Crack patterns in thin films. *J. Mech. Phys. Solids* **2000**, *48*, 1107–1131. [CrossRef]
23. Jin, D.; Wang, J.; You, L.; Ge, D.; Liu, C.; Liu, H.; You, Z. Waste cathode-ray-tube glass powder modified asphalt materials: Preparation and characterization. *J. Clean. Prod.* **2021**, *314*, 127949. [CrossRef]
24. Jin, D.; Boateng, K.A.; Ge, D.; Che, T.; Yin, L.; Harrall, W.; You, Z. A case study of the comparison between rubberized and polymer modified asphalt on heavy traffic pavement in wet and freeze environment. *Case Stud. Constr. Mater.* **2023**, *18*, e01847. [CrossRef]
25. Jin, D.; Yin, L.; Xin, K.; You, Z. Comparison of Asphalt Emulsion-Based Chip Seal and Hot Rubber Asphalt-Based Chip Seal. *Case Stud. Constr. Mater.* **2023**, *18*, e02175. [CrossRef]
26. Jin, D.; Boateng, K.A.; Chen, S.; Xin, K.; You, Z. Comparison of Rubber Asphalt with Polymer Asphalt under Long-Term Aging Conditions in Michigan. *Sustainability* **2023**, *14*, 10987. [CrossRef]
27. Behnia, B.; Buttlar, W.G.; Reis, H. Nondestructive low-temperature cracking characterization of asphalt materials. *J. Mater. Civ. Eng.* **2016**, *29*, 04016294. [CrossRef]
28. Qiu, Y.; Ding, H.; Zheng, P. Toward a better understanding of the low-temperature reversible aging phenomenon in asphalt binder. *Int. J. Pavement Eng.* **2022**, *23*, 240–249. [CrossRef]
29. Luo, H.; Tian, R.; Zheng, Y.; Wang, D.; Chen, S.; Huang, X. Notched specimen cracking test: A novel method to directly measure low-temperature thermal stress of asphalt binder. *Int. J. Pavement Eng.* **2022**, 1–15. [CrossRef]

Communication

Testing and Analysis of Ultra-High Toughness Cementitious Composite-Confined Recycled Aggregate Concrete under Axial Compression Loading

Li He [1], Sheng Peng [2,3,*], Yong-Sheng Jia [2,3], Ying-Kang Yao [2,3,*] and Xiao-Wu Huang [2,3]

1   Hubei Province Key Laboratory of Systems Science in Metallurgical Process, Wuhan 430065, China; emp-heli@hotmail.com
2   State Key Laboratory of Precision Blasting, Jianghan University, Wuhan 430056, China
3   Hubei Key Laboratory of Blasting Engineering, Jianghan University, Wuhan 430056, China
*   Correspondence: pe_sh@sina.com (S.P.); shanxiyao@jhun.edu.cn (Y.-K.Y.)

Abstract: In order to analyze the axial compressive properties of ultra-high-toughness cementitious composite (UHTCC)-confined recycled aggregate concrete (RAC), a batch of UHTCC-confined RAC components was designed and manufactured according to the requirements of GB/T50081-2002 specifications. After analyzing the surface failure phenomenon, load-displacement curves, scanning electron microscope (SEM), and parameter analysis of the specimen, the result shows that UHTCC-confined RAC is an effective confinement method, which can effectively improve the mechanical properties and control the degree of surface failure of RAC structures. Compared with the unconfined specimen, the maximum peak load of the UHTCC confinement layer with a thickness of 10 mm and 20 mm increased by 44.61% and 79.27%, respectively, meeting the requirements of engineering practice. Different fiber mixing amounts have different effects on improving the mechanical performance of RAC structural. The specific rule was steel fiber (SF) > polyvinyl alcohol fiber (PVAF) > polyvinyl alcohol fiber (PEF) > no fiber mixture, and the SF improves the axial compression properties of UHTCC most significantly. When there are strict requirements for improving the mechanical properties of the structure, SF should be added to UHTCC. On the contrary, PVAF should be added to UHTCC.

Keywords: ultra-high toughness cementitious composite (UHTCC); recycled aggregate concrete (RAC); scanning electron microscope (SEM); confinement; engineering practice

Academic Editors: Sérgio Manuel Rodrigues Lopes, Honglei Chang, Li Ai, Feng Guo and Xinxiang Zhang

Received: 3 September 2023
Revised: 22 September 2023
Accepted: 4 October 2023
Published: 6 October 2023

1. Introduction

Recycled aggregate concrete (RAC) is a green building material that has the characteristics of reducing carbon emissions, reducing environmental pollution, and saving nonrenewable natural resources in green buildings [1]. It helps to expand the utilization of building solid waste, reduce natural aggregate consumption, and help solve environmental problems. It is an important way to achieve the sustainability of building structures [2]. As the service duration increases, RAC structures will experience different levels of performance degradation and attenuation because of environmental erosion, natural disasters, and functional changes.

For RAC structures whose performance cannot meet the requirements of the specifications, conventional repair methods such as the enveloped steel jacket repair method [3], the carbon fiber reinforced polymer (CFRP) repair method [4], and the enlarging section method [5] have been adopted and used. However, ESJ has certain limitations in marine green buildings, and it is extremely difficult to meet the requirements for anti-corrosion treatment and chloride ion erosion prevention. CFRP has the advantages of being lightweight, high-strength, convenient construction, good corrosion resistance, and durability. However, the organic binder used in the material-to-material interface constraint will exhibit defects such as aging, poor high temperature resistance, and fire resistance when exposed to extreme conditions such as ultraviolet, humidity, high temperature, or fire for a long time, exposing the shortcomings of CFRP repair methods [6]. The enlarging section method requires sufficient curing time; otherwise, it cannot meet the strength requirements. The mechanization level of the enlarging section method is relatively low, and the labor cost is high. Ultra-high-toughness cementitious composite (UHTCC) is a special cement-based material added with a certain volume of specific chopped fibers, which can effectively reduce the erosion of chloride ions and other harmful examples on the steel inside the structure [7], change the failure mode of concrete single cracks, and have obvious advantages in the collaborative crack resistance of mortar base materials and fiber grids, making it an important repair method in the field of green building reinforcement. UHTCC show significant strain hardening and excellent crack resistance under both tensile and bending loads. There have been many studies on the mechanical properties of beams, columns, joints, and frame structures confined with UHTCC, revealing the reinforcement mechanism and mechanical laws of the confined components and structures [8–13]. Li et al. [14] studied the confinement effect of spraying a 20 mm thin layer of UHTCC at the bottom of a cracked concrete beam, which increased its ultimate bearing capacity by 117.5%. After confinement, the crack width was controlled below 0.1 mm, which was beneficial for improving durability. Kim et al. [15,16] studied the crack's propagation mechanism at the bonding interface between sprayed UHTCC and existing concrete and conducted bending tests on beam components composed of equal-thickness sprayed UHTCC and concrete thin plates. They found that sprayed UHTCC significantly improved the bearing capacity of the composite components and had good crack control ability. Lim et al. [17,18] conducted experiments on UHTCC/concrete T-shaped incision composite beams and found that the UHTCC repair layers have better crack dispersion ability than the repair layers of steel fiber concrete and concrete. Xu Shiniang's team [19–23] conducted bending performance tests on UHTCC-confined concrete composite beams. Zhang et al. [17,24] used UHTCC/concrete composite tensile testing to test the interface bonding tensile behavior between UHTCC and concrete. Zhang et al. [25] found that the number of microcracks in the UHTCC repair layer was related to the length of the UHTCC/concrete stripping interface. Kamada et al. [26] pointed out that in the confinement of defective concrete, decreasing interface roughness can make the performance of the UHTCC repair layer more outstanding. Kim et al. [15,27] studied the bonding behavior between sprayed UHTCC and concrete and found that the effect of sprayed UHTCC and cast-in-place UHTCC was the same. Wang et al. [28] conducted splitting tensile and shear tests on a total of 256 cubic bonding specimens to study the bonding behavior between UHTCC and existing concrete. Some factors, including the interface roughness, compressive strength, interface moisture states of existing concrete, and UHTCC pouring position, were investigated. Jiang et al. [29] focused on the mechanism of strengthening concrete columns with FRP grids/FCC using polyethylene-type FCC as the substrate. The experimental study was conducted on a standard concrete cylinder strengthened by an FRP grid/FCC, where new FCC material served as the matrix. The testing variables were plain concrete strength and different textile grids, i.e., basalt fiber reinforced polymer (BFRP) and CFRP grid. The uniaxial compressive performance was studied. The confinement of green building structures is essentially the confinement of one side of the interface between UHTCC and RAC, rather than wrapping the RAC to form a complete confinement mode. In order to provide a basis for the load design of the UHTCC-confined RAC interface under axial compression load, it is extremely important to study the bonding behavior between the UHTCC and RAC interface by referring to the research results of the UHTCC and concrete bonding behavior [30]. Therefore, it is still necessary to supply and complete the research on the axial compressive mechanical behavior of UHTCC-confined RAC.

The factors that affect the interface of UHTCC-confined RAC include two parts: UHTCC and RAC. The research on the influential mechanism of green building structures that meet the requirements of design specifications was relatively mature, so the research on the impact mechanism of UHTCC was mainly focused on. Taking full account of the influence of UHTCC fiber type and reinforcement layer thickness, axial compression tests and stress analysis were carried out on UHTCC-confined RAC to study its obvious failure phenomenon, load displacement curve, scanning electron microscope (SEM), and parameter effects. The dispersion and existence states of different fiber types in the matrix were obtained by SEM, revealing the influence of the UHTCC protective layer and fiber type on the compression performance of RAC.

2. Test Overview

2.1. Materials

During preparation, both the Hobart and cement paste mixers were used for UHTCC and RAC slurry mixing. In this study, the PI 42.5 reference Portland cement according to Chinese standard (GB8076-2016) [31], tap water, low calcium fly ash, medium sand (the particle grading conforms to Class II grading zone), gravel (produced by Hebei Yueshan Environmental Protection Technology Co., Ltd.), polycarboxylate superplasticizer (VIVID-651, produced by Guizhou Hengfan New Technology Development Co., Ltd.), and recycled coarse aggregate (produced by Hangzhou Ruichen Building Materials Co., Ltd.) were used to prepare RAC paste. The mix proportion of RAC is cement (kg/cm^3):water (kg/cm^3):medium sand (kg/cm^3):gravel (kg/cm^3):low calcium fly ash (kg/cm^3):polycarboxylate superplasticizer (kg/cm^3):recycled coarse aggregate (kg/cm^3) = 1.0:0.89:2.29:5.26:0.74:0.04:2.29. The density is 2400 kg/cm^3, the water-binder ratio is 0.89, the replacement rate of recycled coarse aggregate is 50%, the strength grade of RAC is C40, and the measured value of standard cubic axial compressive strength is 39.2 MPa.

While for the UHTCC paste, the binder materials include PI 52.5 Portland cement according to Chinese standard (GB175-2017) [32], tap water, medium sand (the particle grading conforms to Class II grading zone), polycarboxylate superplasticizer (VIVID-651, produced by Guizhou Hengfan New Technology Development Co., Ltd.), and low calcium fly ash (produced by Tianjin Zhucheng New Material Technology Co., Ltd.) were used instead. The mix proportion of UHTCC is cement (kg/cm^3), low calcium fly ash (kg/cm^3), medium sand (kg/cm^3), water (kg/cm^3), and polycarboxylate superplasticizer (kg/cm^3) = 1.0:4.0:1.0:1.1:0.04. The density is 2500 kg/cm^3, and the water-binder ratio is 1.1. Polyvinyl alcohol fiber (PVAF), polyethylene fiber (PEF), and steel fiber (SF) that meet Chinese standards were used and shown in Figure 1.

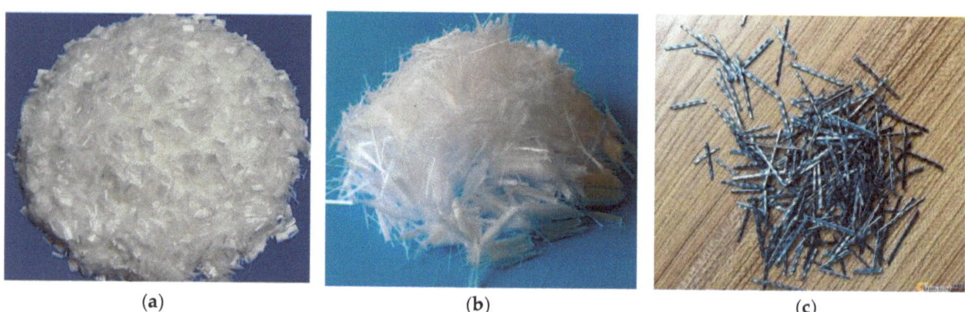

(a)   (b)   (c)

Figure 1. Appearance and shape of three fibers. (**a**) Polyvinyl alcohol fiber (PVAF); (**b**) Polyethylene fiber (PEF); (**c**) Steel fiber (SF).

Moreover, add 2.5% volume content of fibers to the UHTCC slurry to ensure fluidity [33]. The length, diameter, length-width ratio, density, tensile strength, and elastic modulus of PVAF are 12 mm, 0.041 mm, 387.1, 1.3 g·cm^3, 1560 MPa, and 41 GPa, respectively. The length, diameter, length-width ratio, density, tensile strength, and elastic modulus of PEF are 0.0068 mm, 0.5 mm, 324.8, 0.941~0.965 g·cm^3, 21~38 MPa, and 0.84~0.95 GPa, respectively. The length, diameter, length-width ratio, density, tensile strength, and elastic modulus of SF are 30 mm, 0.32 mm, 75, 7.85 g·cm^3, 86.5 MPa, and 80~90 GPa, respectively.

2.2. Specimens Design

The size of the specimen in the axial compression strength tests is 150 mm × 150 mm × 150 mm according to Chinese standard (GB/T50081-2002) [34], including 7 sets of 3 specimens in each group, totaling 21 specimens. The experimental variables include the reinforcement layer thickness and fiber type of UHTCC. The reinforcement layer thicknesses of UHTCC are 0, 10 mm, and 20 mm. The fiber types of UHTCC include PVAF, PEF, and SF. The numbering and configuration details of the test specimens are shown in Table 1.

Table 1. Specimen numbering and configuration details.

Specimen No.	Recycled Concrete Strength	Protective Layer Thickness/mm	Fiber Type	Peak Load/kN
C40	C40	0	-	
	C40	0	-	401.92
	C40	0	-	
C40-PVAF-10	C40	10	PVAF	
	C40	10	PVAF	581.23
	C40	10	PVAF	
C40-PVAF-20	C40	20	PVAF	
	C40	20	PVAF	720.53
	C40	20	PVAF	
C40-PEF-10	C40	10	PEF	
	C40	10	PEF	485.88
	C40	10	PEF	
C40-PEF-20	C40	20	PEF	
	C40	20	PEF	572.33
	C40	20	PEF	
C40-SF-10	C40	10	SF	
	C40	10	SF	519.82
	C40	10	SF	
C40-SF-20	C40	20	SF	
	C40	20	SF	581.53
	C40	20	SF	

2.3. Test Steps and Process

As shown in the loading process in Figure 2, the axial load is provided by the WAW-E series microcomputer-controlled electro-hydraulic servo universal testing machine (Jinan Zhongluchang Testing Machine Manufacturing Co., Ltd.), which meets the standards of GB/T16826-2008, GB/T228-2010, and GB/T7314-2005 [35–37]. The loading rate should be controlled at 0.5 MPa/s until it is destroyed. The specimens from top to bottom are UHTCC, interface layers, and RAC. During the test, it is necessary to apply a uniformly thin layer of Vaseline to the upper and lower surfaces of the specimen. Collect damaged specimens, polish the surface, mark, and scan with SEM to obtain the interface microstructure of the fragments.

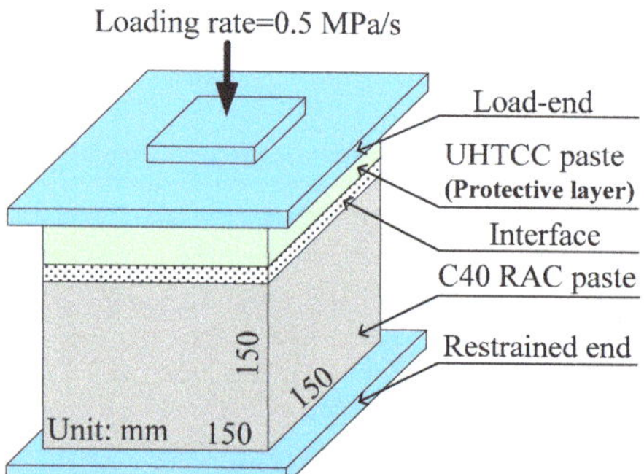

Figure 2. Loading process of the axial compression test.

3. Analysis of Test Results

3.1. Failure Analysis

Whether the fiber type and reinforcement layer thickness of UHTCC enhance the resistance levels of RAC to external forces needs to be determined based on the failure mode and failure rule of UHTCC-confined RAC under axial compression. As shown in the apparent failure phenomenon in Figure 3, the RAC failure mode of C40 strength degree has no special characteristics under axial compression, only forming vertical cracks and crushing at both ends. The failure degree of the specimen C40 is the highest, indicating that the UHTCC repair method is a good method to improve mechanical properties. For the degree of apparent failure, specimens C40-PVAF-10 < C40-PVAF-20, C40-PEF-10 < C40-PEF-20, and C40-SF-10 < C40-SF-20 indicate that, when the thickness reinforcement layer of UHTCC reaches 20 mm, the bearing properties of the specimen will be improved. The appearance of specimens C40-PVAF-10, C40-PVAF-20, C40-PEF-10, and C40-PEF-20 showed significant diagonal cracking, with specimen C40-PVAF-20 being the most prominent, indicating that both PVAF and PEF are beneficial for enhancing toughness and that PVAF is superior to PEF. The apparent specimens C40, C40-SF-10, and C40-SF-20 were mainly showed vertical penetrating cracking. From the perspective of cracking degree, specimens C40 > C40-SF-10 > and C40-SF-20 indicate that when the reinforcement layer thickness of UHTCC is 20 mm, the bearing properties of the structure can be effectively solved.

3.2. Micro-Structure Analysis

In order to analyze the micro-structure of UHTCC-confined RAC, a 1:80 μm measurement range was selected to conduct SEM of the specimen fragments after apparent failure identification. The internal defects of RAC can easily cause discontinuity or honeycomb shapes, forming discontinuous sponge-like distributions (as shown in Figure 4g), which is also the reason for the strong brittleness of RAC. The addition of PVAF can greatly enhance toughness; the cracks formed were suppressed in both width and length, and the number was also controlled. Moreover, the direction of the cracks was uncontrolled, and the formation of a mesh support structure was formed (as shown in Figure 4a,b), which significantly enhanced the bearing properties of the specimen. The addition of PEF basically maintained a very similar effect to the addition of PVAF, but the difference in slenderness ratio between the two resulted in the inability of PEF addition to form a stable network support structure, only forming a membrane structure (as shown in Figure 4c,d), which reduced the toughness in terms of reinforcement. The addition of SF can significantly increase frictional resistance and mechanical interaction (as shown in Figure 4e,f) and has a good inhibiting effect on crack development. Its action mechanism is similar to that between reinforcement bars and concrete, but it is extremely difficult to change the trend of crack development, resulting in a very similar apparent phenomenon in specimens C40, C40-SF-10, and C40-SF-20.

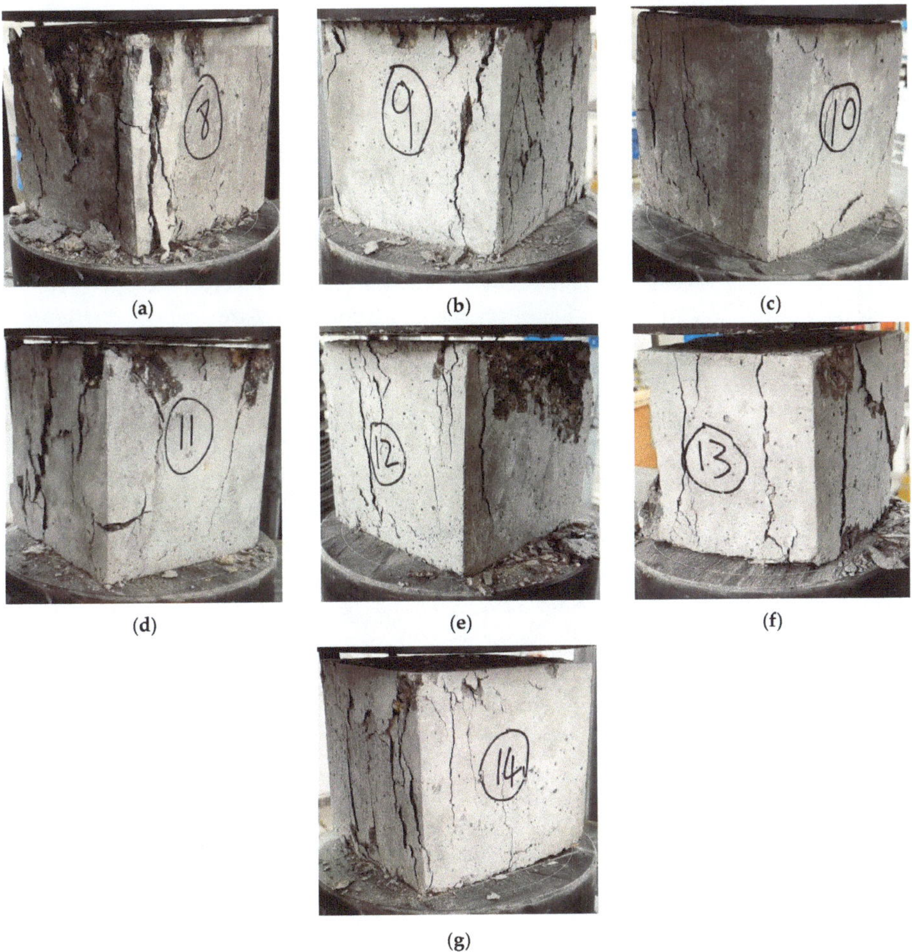

Figure 3. Diagram of failure mode. (**a**) Specimen C40; (**b**) Specimen C40-PVAF-10; (**c**) Specimen C40-PVAF-20; (**d**) Specimen C40-PEF-10; (**e**) Specimen C40-PEF-20; (**f**) Specimen C40-SF-10; (**g**) Specimen C40-SF-20.

3.3. Mechanical Properties

Figure 5 shows the load-displacement curves of each specimen under an axial compression load. In terms of peak load, specimen C40 is the smallest, indicating that the UHTCC repair method is an effective reinforcement method. Comparing the load-displacement curves, it was found that the minimum displacement was required for the descending section of specimen C40, followed by specimens C40-SF-10 and C40-SF-20. There was no significant difference between specimens C40-PVAF-10, C40-PVAF-20, C40-PEF-10, and C40-PEF-20, indicating that the addition of PVAF and PEF had the most significant effect on toughness enhancement, followed by SF. Therefore, priority should be given to adding PVAF or PEF fiber materials to the UHTCC slurry.

3.4. Influence of UHTCC Protective Layers

Figure 6 shows the histograms of the peak load of each specimen under different UHTCC reinforcement layers. In this paper, the UHTCC reinforcement layers are set to 0, 10 mm, and 20 mm. When the UHTCC reinforcement layer is 0, the peak load is the smallest, followed by the specimen with a UHTCC reinforcement layer of 10 mm. The specimen with a UHTCC reinforcement layer thickness of 20 mm has the highest peak load, indicating that the greater the thickness of the UHTCC reinforcement layer, the better the improvement in mechanical properties. When the reinforcement layer thickness of UHTCC is 20 mm, the variation pattern of peak load is specimens C40-SF-20 > C40-PVAF-20 > C40-PEF-20. When the reinforcement layer thickness of UHTCC is 10 mm, the variation pattern of peak load was specimens C40-SF-10 > C40-PVAF-10 > C40-PEF-10. Compared with specimen C40, the peak loads of specimens C40-SF-20, C40-PVAF-20, and C40-PEF-20 increased by 79.27%, 44.69%, and 42.40%, respectively; the peak loads of specimens C40-SF-10, C40-PVAF-10, and C40-PEF-10 increased by 44.61%, 29.33%, and 20.89%, respectively. The addition of SF to UHTCC can significantly increase the axial compression mechanical properties of RAC. The thickness reinforcement layers of 10 mm and 20 mm can increase the axial force of the structure, which meets the needs of engineering practice. The order of the improvement effect on structural mechanics properties is PVAF > SF > PEF > no addition fiber.

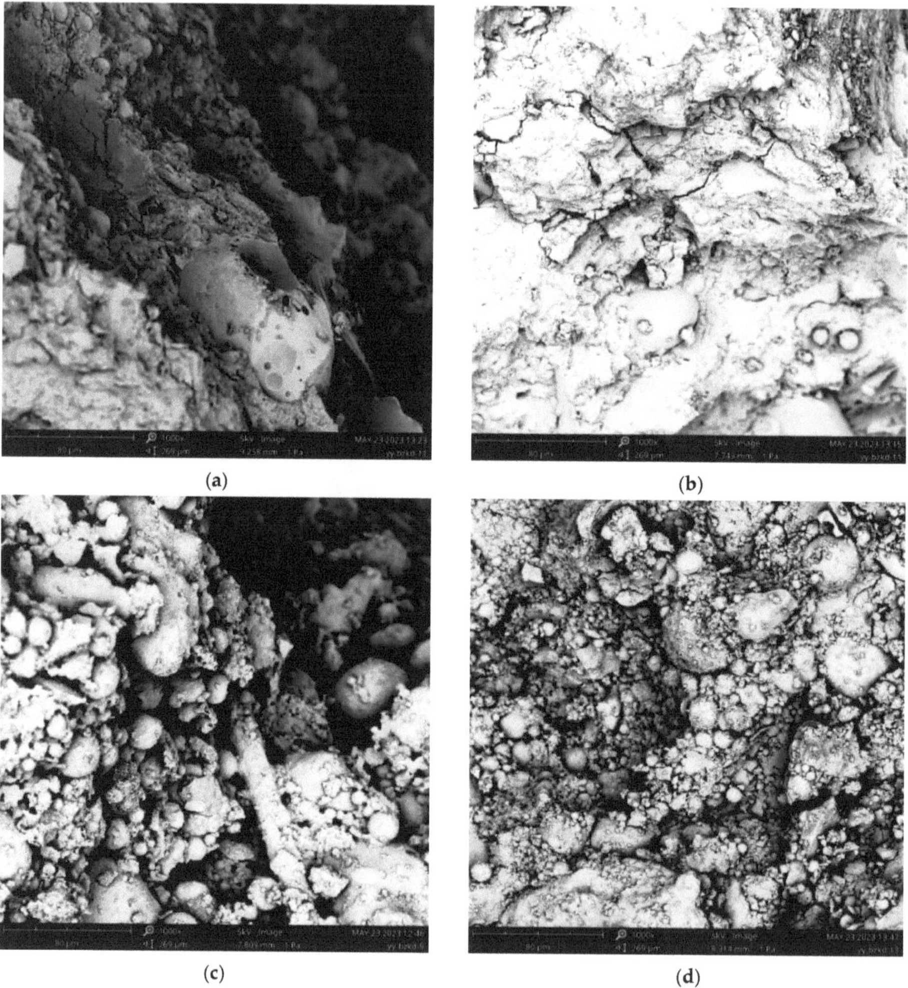

(a)

(b)

(c)

(d)

Figure 4. *Cont.*

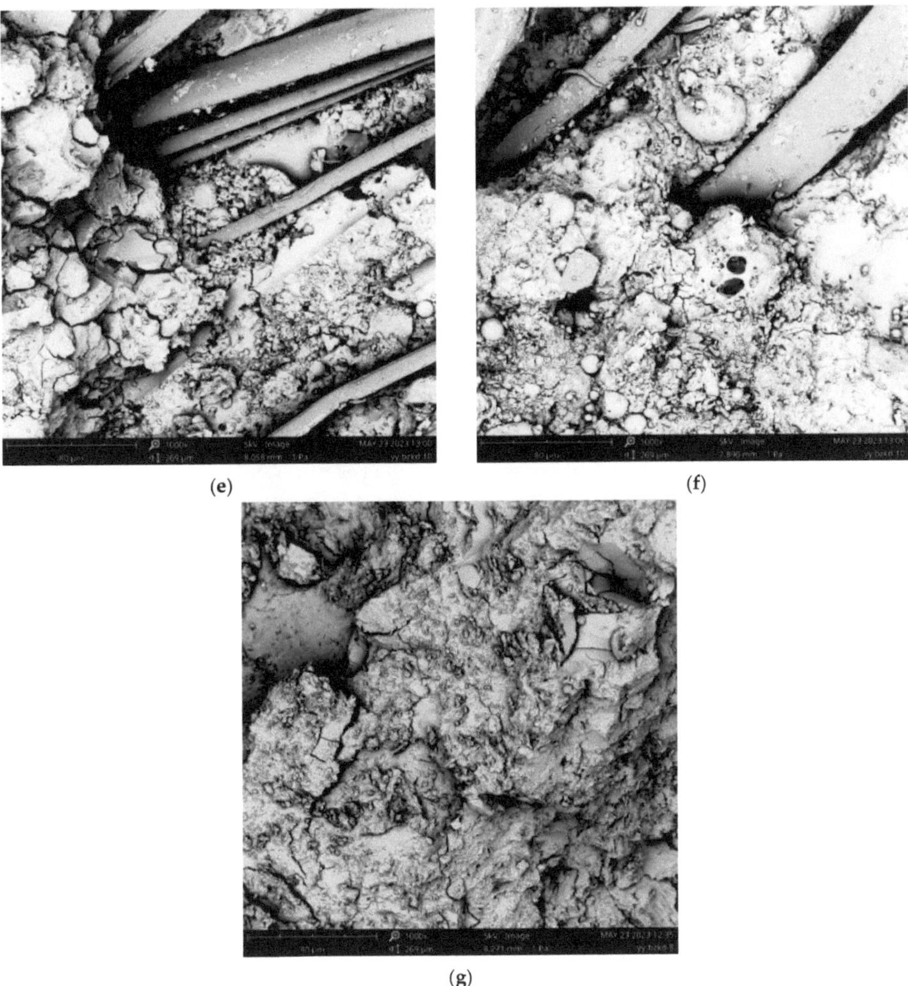

Figure 4. SEM of microstructure with three-type fibers.; (**a**) Specimen C40-PVAF-10; (**b**) Specimen C40-PVAF-20; (**c**) Specimen C40-PEF-10; (**d**) Specimen C40-PEF-20; (**e**) Specimen C40-SF-10; (**f**) Specimen C40-SF-20; (**g**) Specimen C40.

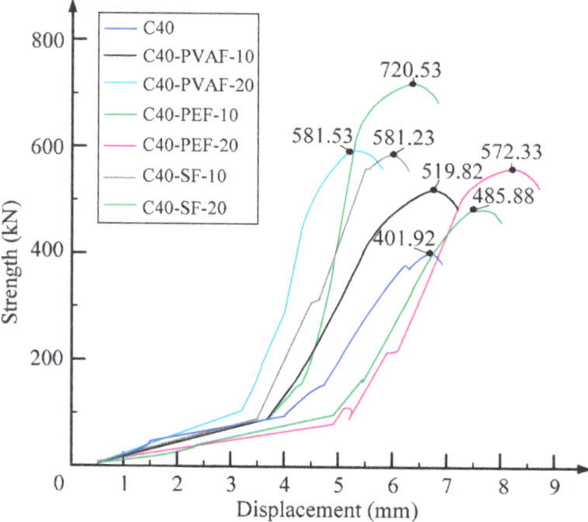

Figure 5. Load-displacement curves with three-type fibers.

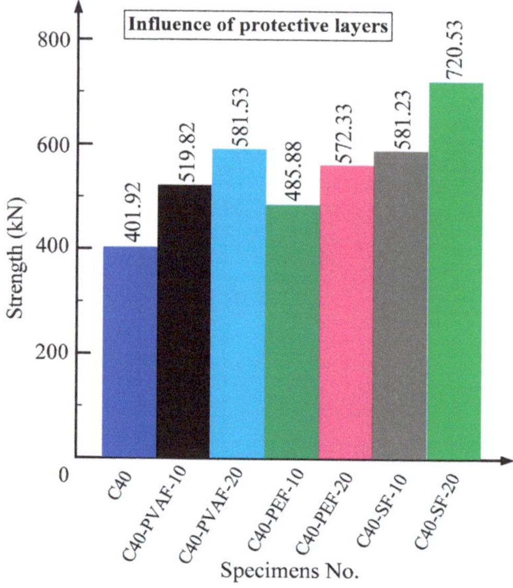

Figure 6. Influence of UHTCC protective layers.

3.5. Influence of UHTCC Fiber Types

Figure 7 shows the histograms of the peak load of each specimen under different types of UHTCC fibers. According to the different reinforcement layers of UHTCC, it can be divided into two groups: 10 mm and 20 mm. Compared with specimen C40-PEF-10, the peak load of specimens C40-SF-10 and C40-PVAF-10 increased by 19.62% and 6.99%, respectively. Compared with specimen C40-PEF-20, the peak loads of specimens C40-SF-20 and C40-PVAF-20 increased by 25.89% and 1.61%, respectively, indicating that the axial compression behavior of UHTCC with the addition of SF has the most significant improvement, but the improvement amplitude does not show a linear relationship with the increase in reinforcement layer thickness of the UHTCC. It is recommended that the UHTCC reinforcement layer have a thickness of 10 mm. If there is a high requirement for improving mechanical properties, it is recommended to use the addition of SF to UHTCC; otherwise, please use the addition of PVAF to UHTCC.

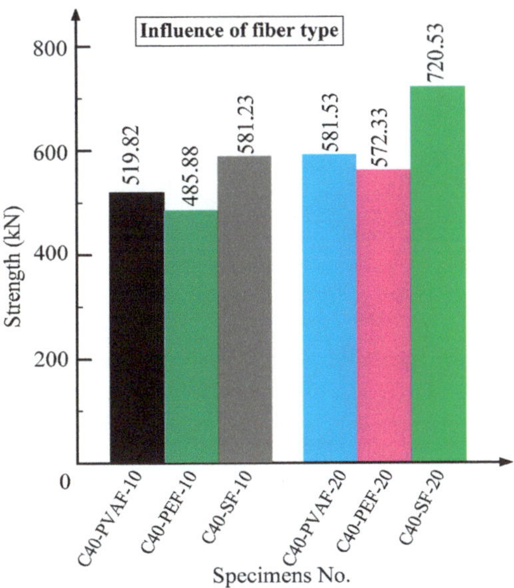

Figure 7. Influence of UHTCC fiber types.

4. Summary and Conclusions

The influence of reinforcement layer thickness and fiber type on UHTCC-confined RAC cannot be ignored. Through a series of axial compression failure tests, the apparent failure phenomenon, load-displacement curves, SEM, and parameter effects of the specimens were obtained. The conclusions are as follows:

1. UHTCC reinforcement is an effective repair method. The mechanical properties of UHTCC-confined specimens have significantly improved, and the degree of apparent damage has been effectively controlled;

2. The variation pattern of the peak load is as follows: specimens C40-SF-20 > C40-SF20 > C40-PVAF-10 > C40-PEF-20 > C40-SF-10 > C40-PEF-10 > C40, indicating that the reinforcement layer thickness of UHTCC is 10 mm and 20 mm, respectively. This can effectively solve the bearing performance problem of RAC;

3. Compared with specimen C40, the peak loads of specimens C40-SF-20, C40-PVAF-20, and C40-PEF-20 increased by 79.27%, 44.69%, and 42.40%, respectively; the peak loads of specimens C40-SF-10, C40-PVAF-10, and C40-PEF-10 increased by 44.61%, 29.33%, and 20.89%, respectively. The addition of SF to UHTCC can significantly increase the axial compression mechanical properties of RAC. The order of the improvement effect on structural mechanics properties is PVAF > SF > PEF > no addition fiber;

4. Compared with specimen C40-PEF-20, the peak loads of specimens C40-SF-20 and C40-PVAF-20 increased by 25.89% and 1.61%, respectively, indicating that the axial compression behavior of UHTCC with the addition of SF has the most significant improvement, but the improvement amplitude does not show a linear relationship with the increase in reinforcement layer thickness of the UHTCC. It is recommended that a UHTCC reinforcement layer thickness of 10 mm;

5. If there is a high requirement for improving mechanical properties, it is recommended to use the addition of SF to UHTCC; otherwise, please use the addition of PVAF to UHTCC. Compared with PVAF and PEF. For improving the mechanical properties and 10 mm confinement layer, please add SF to UHTCC.

Author Contributions: S.P.: data curation, formal analysis, investigation, methodology, project administration and writing—review and editing; L.H.: data curation, formal analysis, investigation, and writing—review and editing; Y.-S.J., Y.-K.Y. and X.-W.H.: formal analysis, funding acquisition, investigation, and methodology. All authors have read and agreed to the published version of the manuscript.

Funding: This research was funded by [National Natural Science Foundation of China] grant number [51904210] And The APC was funded by [51904210], Supported by [State Key Laboratory of Precision Blasting and Hubei Key Laboratory of Blasting Engineering of China] grant number [PB-SKL2022D05], [Natural Science Foundation of Hubei Province] grant number [2022CFB662], [Key Laboratory of Impact and Safety Engineering (Ningbo University), Ministry of Education of China] grant number [CJ202306], [Wuhan Knowledge Innovation Special Basic Research Project] grant number [2022020801010371], and heir support is gratefully acknowledged.

Data Availability Statement: Some or all data, models, or code that support the findings of this study are available from the corresponding authors upon reasonable request.

Conflicts of Interest: The authors declare no conflict of interest.

References

1. Liang, G.W.; Yao, W.; Wei, Y.Q. A green ultra-high performance geopolymer concrete containing recycled fine aggregate: Mechanical properties, freeze-thaw resistance and microstructure. *Sci. Total Environ.* **2023**, *895*, 165090. [CrossRef] [PubMed]
2. Miguel, F.; Brito, J.D.; Silva, R.V. Durability-related performance of recycled aggregate concrete containing alkali-activated municipal solid waste incinerator bottom ash. *Constr. Build. Mater.* **2023**, *397*, 132415. [CrossRef]
3. Zhao, X.Y.; Chen, J.X.; Chen, G.M.; Xu, J.J.; Zhang, L.W. Prediction of ultimate condition of FRP-confined recycled aggregate concrete using a hybrid boosting model enriched with tabular generative adversarial networks. *Thin-Wall. Struct.* **2023**, *182*, 110318. [CrossRef]
4. He, A.; Cai, J.; Chen, Q.J.; Liu, X.; Xue, H.; Yu, C. Axial compressive behaviour of steel-jacket retrofitted RC columns with recycled aggregate concrete. *Constr. Build. Mater.* **2017**, *141*, 501–516. [CrossRef]
5. Han, Q.; Yuan, W.Y.; Ozbakkaloglu, T.; Bai, Y.L.; Du, X.L. Compressive behavior for recycled aggregate concrete confined with recycled polyethylene naphthalate/terephthalate composites. *Constr. Build. Mater.* **2020**, *261*, 120498. [CrossRef]
6. Wang, Z.H. Research review of concrete columns strengthened with fiber reinforced polymer (FRP). *J. Nanjing Inst. Technol.* **2012**, *10*, 19–24. (In Chinese)
7. Li, V.C. *Advances in ECC Research*; ACI Special Publication on Concrete: Material Science to Applications; American Concrete Institute: Indiana, IN, USA, 2002; Volume SP (206-223), pp. 373–400.
8. Zheng, Y.Z.; Wang, W.W. Experimental research on flexural behavior of RC beams strengthened with FRP grid-UHDCC composite. *China Civ. Eng. J.* **2017**, *50*, 23–32. (In Chinese)
9. Zhu, Z.F.; Wang, W.W. Experimental study on mechanical behavior of circular reinforced concrete columns strengthened with FRP textile and ECC. *J. Southeast Univ.* **2016**, *46*, 1082–1087.
10. Zhu, Z.F.; Wang, W.W. Experiment on the uniaxial tensile mechanical behavior of basalt grid reinforced engineered cementitious composites and its constitutive model. *Acta Mater. Compos. Sin.* **2017**, *34*, 2367–2374.
11. Li, Z.; Li, Y. Axial compression performance study on basalt braided mesh composite ECC reinforced concrete cylinder. *Build. Struct.* **2017**, *47*, 40–44.
12. Zhu, Z.F.; Wang, W.W. Constitutive model of basalt fiber reinforced polymer grid reinforced engineered cement composite under cyclic loading. *J. Nanjing Tech Univ.* **2017**, *3*, 44–50. (In Chinese)
13. Mathews, M.E.; Anand, N.; Kodur, V.K.; Arulraj, P. The bond strength of self-compacting concrete exposed to elevated temperature. *Proc. Inst. Civ. Eng. Struct. Build.* **2021**, *174*, 804–821. [CrossRef]
14. Li, W.P.; Mu, F.J.; Wang, J.Y.; Ding, H.Y.; Xu, S.L. Experimental study on mechanical performance of cracked plain concrete beams retrofitted by sprayed ultra high toughness cementitious composites. *World Earthq. Eng.* **2018**, *34*, 166–172.
15. Kim, Y.Y.; Kong, H.J.; Li, V.C. Design of engineered cementitious composite suitable for wet-mixture shotcreting. *ACI Mater. J.* **2003**, *100*, 511–518.
16. Kim, Y.Y.; Li, V.C. Mechanical performance of sprayed engineered cementitious composite suitable for wet-mixture shotcreting process for repair applications. *ACI Mater. J.* **2004**, *101*, 42–49.
17. Lim, Y.M.; Li, V.C. Durable repair of aged infrastructures using trapping mechanism of engineered cementitious composites. *Cem. Concr. Compos.* **1997**, *19*, 373–385. [CrossRef]
18. Lim, Y.M.; Li, V.C. Trapping mechanism of interface crack at concrete/engineered cementitious composite bimaterial interface. Damage and Failure of Interface 1st International Conference on Damage and Failure of Interfaces. In Proceedings of the First International Conference on Damage and Failure of Interfaces—DFI-1, Vienna, Austria, 22–24 September 1997; pp. 405–409.
19. Li, H.D.; Leung, C.K.Y.; Xu, S.L.; Cao, Q. Potential use of strain hardening ECC in permanent formwork with small scale flexural beams. *J. Wuhan Univ. Technol.* **2009**, *24*, 482–487. [CrossRef]

20. Xu, S.; Xie, H.; Li, Q.; Hong, C.; Liu, J.; Wang, Q. Bending performance of RC slabs strengthened by CFRP sheets using sprayed UHTCC as bonding layer: Experimental study and theoretical analysis. *Eng. Struct.* **2023**, *292*, 116575. [CrossRef]
21. Zhang, X.F.; Xu, S.L.; Li, H.D. Theoretical analysis of flexural performance of plain concrete composite beams strengthened with ultrahigh toughness cementitious composite. *China Civ. Eng. J.* **2010**, *43*, 51–62.
22. Xu, S.L.; Wang, N.; Li, Q.H. Experimental study on the flexural performance of concrete beam strengthened with ultra high toughness cementitious composites. *China Civ. Eng. J.* **2010**, *5*, 17–22.
23. Wang, B.; Xu, S.L.; Liu, F. Evaluation of tensile bonding strength between UHTCC repair materials and concrete substrate. *Constr. Build. Mater.* **2016**, *112*, 595–606. [CrossRef]
24. Zhang, J.; Li, V.C.; Nowak, A.; Wang, S. Introducing ductile strip for durability enhancement of concrete slabs. *J. Mater. Civ. Eng.* **2002**, *14*, 253–261. [CrossRef]
25. Zhang, J.; Li, V.C. Monotonic and fatigue performance in bending of fiber-reinforced engineered cementitious composite in overlay system. *Cem. Concr. Compos.* **2002**, *32*, 415–423. [CrossRef]
26. Kamada, T.; Li, V.C. The effects of surface preparation on the fracture behavior of ECC/concrete repair system. *Cem. Concr. Compos.* **2000**, *22*, 423–431. [CrossRef]
27. Li, V.C. Durable overlay systems with engineered cementitious composites (ECC). *Int. J. Restor. Build. Monum.* **2003**, *9*, 215–234.
28. Wang, N.; Xu, S.L. Bonding performance between ultra high toughness cementitious composites and existing concrete. *J. Build. Mater.* **2011**, *14*, 317–325.
29. Jiang, J.F.; Sui, K. Experimental study of compression performance of concrete cylinder strengthened by textile reinforced engineering cement composites. *Acta Mater. Compos. Sin.* **2019**, *36*, 1957–1967.
30. He, S.; Li, L.; Xiong, Z.; Zhang, H.; Zheng, J.; Su, Y.; Liu, F. Effects of ring-type and straight steel fibres on the compressive performance of rubber-recycled aggregate concrete. *J. Build. Eng.* **2023**, *76*, 107148.
31. GB8076-2016; Concrete Admixtures. Standardization Administration of China: Beijing, China, 2016.
32. GB175-2017; Common Portland Cement. Standardization Administration of China: Beijing, China, 2017.
33. Sabireen; Butt, F.; Ahmad, A.; Ullah, K.; Zaid, O.; Shah, H.A.; Kamal, T. Mechanical performance of fiber-reinforced concrete and functionally graded concrete with natural and recycled aggregates. *Ain Shams Eng. J.* **2023**, *14*, 102121. [CrossRef]
34. GB/T50081-2002; Standard for Test Method of Mechanical Properties on Ordinary Concrete. Standardization Administration of China: Beijing, China, 2002.
35. GB/T16826-2008; Electro-Hydraulic Servo Universal Testing Machines. Standardization Administration of China: Beijing, China, 2008.
36. GB/T228-2010; Metallic Materials-Tensile Testing-Method of Test at Ambient Temperature. Standardization Administration of China: Beijing, China, 2010.
37. GB/T7314-2005; Metallic Materials-Compression Testing at Ambient Temperature. Standardization Administration of China: Beijing, China, 2005.

materials

Article

Nondestructive Detection and Early Warning of Pavement Surface Icing Based on Meteorological Information

Jilu Li [1], Hua Ma [2], Wei Shi [3], Yiqiu Tan [1,*], Huining Xu [1,*] , Bin Zheng [1] and Jie Liu [2]

[1] School of Transportation Science and Engineering, Harbin Institute of Technology, Harbin 150090, China
[2] Xingtai Pavement & Bridge Construction Group Co., Ltd., Xingtai 054000, China
[3] Heilongjiang Transportation Investment Group Co., Ltd., Harbin 150000, China
* Correspondence: tanyiqiu@hit.edu.cn (Y.T.); xuhn@hit.edu.cn (H.X.)

Abstract: Monitoring and warning of ice on pavement surfaces are effective means to improve traffic safety in winter. In this study, a high-precision piezoelectric sensor was developed to monitor pavement surface conditions. The effects of the pavement surface temperature, water depth, and wind speed on pavement icing time were investigated. Then, on the basis of these effects, an early warning model of pavement icing was proposed using an artificial neural network. The results showed that the sensor could detect ice or water on the pavement surface. The measurement accuracy and reliability of the sensor were verified under long-term vehicle load, temperature load, and harsh natural environment using test data. Moreover, pavement temperature, water depth, and wind speed had a significant nonlinear effect on the pavement icing time. The effect of the pavement surface temperature on icing conditions was maximal, followed by the effect of the water depth. The effect of the wind speed was moderate. The model with a learning rate of 0.7 and five hidden units had the best prediction effect on pavement icing. The prediction accuracy of the early warning model exceeded 90%, permitting nondestructive and rapid detection of pavement icing based on meteorological information.

Keywords: icy pavement; intelligent monitoring; meteorological parameter; early warning

Citation: Li, J.; Ma, H.; Shi, W.; Tan, Y.; Xu, H.; Zheng, B.; Liu, J. Nondestructive Detection and Early Warning of Pavement Surface Icing Based on Meteorological Information. *Materials* **2023**, *16*, 6539. https://doi.org/10.3390/ma16196539

Academic Editor: Changho Lee

Received: 23 August 2023
Revised: 19 September 2023
Accepted: 30 September 2023
Published: 3 October 2023

1. Introduction

Ice refers to a pollutant on a pavement surface formed by rain condensation, melting snow, or humid air in low-temperature environments. Ice adheres to pavement surfaces and masks the surfaces' texture [1], thereby causing a rapid reduction in pavement anti-sliding characteristics [2]. Accidents on icy pavements have a high occurrence probability and serious consequences. They also gravely threaten the personal safety of drivers [3,4]. The presence of ice on pavements should be accurately detected, and timely warning must be provided to reduce the occurrence of traffic accidents in winter.

The accurate detection of pavement ice may contribute considerably to traffic control decision-making and improve traffic safety [5,6]. Accordingly, various ice sensors with different technologies have been developed to detect pavement conditions. In accordance with the detection method, sensors can be divided into destructive sensing and nondestructive sensing. Destructive detection sensors are installed on pavements to detect ice (e.g., sonic wave [7–9], resistance [10], capacitance [11], and vibration [12]). The limitations of the aforementioned sensors, such as sluggish response times and poor durability, prevent their application in pavement engineering. Additionally, nondestructive detection sensors, such as infrared [13,14] and optical fiber [15], are installed above the pavement surface. Although the accuracy of nondestructive sensing technology is high under normal operations, the interference of pavement lights and installation height have potential influence on the technology's detection accuracy, making stable and accurate detection results difficult to obtain. In addition, the current sensors are relatively expensive. China's extensive transportation network makes it difficult to use these sensors for coverage applications. Therefore, ice sensors for pavement engineering should be developed considering cost, durability, and accuracy.

Meteorological factors are the main cause of pavement icing. The effects of various factors on the icing state have been studied. Zhang et al. simulated the icing environment to study the nonlinear formation of ice on a cold surface under the action of meteorological factors. The test results showed that the icing time increased with the decrease in temperature and wind speed [16]. Xu et al. measured the icing time of a pavement and established an empirical relationship among pavement temperature, water freezing point, and icing time [17]. Samodurova analyzed the influence of various factors on pavement icing and proposed a linear discriminant forecast function with crucial factors: temperature and precipitation [18,19].

Having clarified the influence of meteorological factors on icing, researchers have also attempted to achieve nondestructive detection and warning of pavement icing using meteorological factors. On the basis of the pavement meteorological data collected by the Pavement Weather Information System, the Swedish Transport Administration studied the effects of pavement temperature and precipitation on the pavement icing state and developed the pavement icing expert analysis system [20,21]. The relevant departments in Germany also attach great importance to pavement weather forecasts. A pavement management department analyzed the influence of temperature and wind speed on pavement icing and proposed a pavement icing warning method that considers the meteorological information in the past 24 h [22]. The National Weather Service in the United States established a pavement information monitoring system. The system can predict the thickness of pavement ice accretion on the basis of the variations in several factors (e.g., air temperature, wind speed, and precipitation rate) [23]. Korotenko investigated the relationship among temperature, water, and pavement icing; a second-order diffusion equation with empirically parameterized flux terms was proposed for the numerical icing forecasting system in Northern Europe and North America [24]. Meanwhile, Martorina and Loglisci conducted thermal mapping of main pavements in Piemonte to infer the pavement temperature along all pavement networks. The researchers proposed a small-scale pavement ice forecasting system in accordance with the variation in air temperature, dew point, wind speed, and cloud amount in the past three hours [25]. Alexander summarized and predicted the icing-sensitive areas of the Danish highway network using observed data (pavement temperature, air temperature, and dew point temperature) from a meteorological station from 2003 to 2007. The researchers suggested improving the quality and accuracy of pavement icing prediction through thermal mapping measurements [26].

Moreover, an accurate, nondestructive detection and early warning model provides essential benefits. Ye et al. showed that the ice early warning system developed by the California Department of Transportation has economized at least USD 1.7 million and reduced the occurrence of accidents by 18% [27]. The models above can predict pavement icing successfully through the nonlinear regression of the historical data of meteorological factors. However, this type of early warning entails some problems. In low-temperature environments, the water on pavements takes some time to completely turn into ice. The abovementioned models are complicated because excessive factors are considered. Too many model input parameters lead to the need for a large calculation capacity and long warning times, which cannot meet the needs of rapid warning of pavement icing. Furthermore, the models exhibit strong temporal and spatial characteristics, which limit their application. With the continuous change in the global climate, the applicability of these models decreases, causing a decline in early warning accuracy. Thus, fast-response, self-adaptive early warning models of pavement icing are needed.

Due to a variety of factors, China's road management still lacks an effective nondestructive detection and early warning system for pavement icing. In order to eliminate the negative effects of ice on highways in Hebei Province, this paper preliminarily realizes the rapid, nondestructive detection and precise, accurate early warning of pavement icing for the meteorological conditions of Hebei Province. Specifically, this study proposes a pavement surface icing sensor that enables the accurate measurement of icing thickness through the piezoelectric properties of the material. The effects of key meteorological factors on pavement icing were clarified, and an adaptive learning pavement icing warning model was developed to nondestructive detect pavement icing by inputting meteorological parameters. The study provides novel insights into and a technical reference for the safe and efficient operation of pavement infrastructure.

2. Sensor and Method

2.1. Sensor Development

Accurate acquisition of pavement icing information is the key to improving traffic safety in the winter. Since sensors have a high failure rate and low accuracy when used to obtain pavement icing information because of the hostile natural environment of the pavement, manual observation is still the primary method used to obtain this information today. This observational method relies on the empirical assessment of pavement maintenance workers, which might be very subjective and prone to inaccuracy. In order to increase the precision and automation of ice detection, this paper proposes a durable ice sensor for acquiring icing condition information.

2.1.1. Sensor Design

1.   Sensor Principle

Through the change in the overall resonant frequency of the sensing element and the attached material, the sensor determines the thickness of the material on the surface of the sensing element. When the sensor is in operation, a stimulus electrical signal is applied to the input electrode of the sensing element in a certain step, and periodically increases, causing the piezoelectric ceramic to vibrate. The output electrode of the sensor element produces a voltage signal that contains information about the vibration frequency as a result of the inverse piezoelectric effect. The output electrical signal will be greatest when the frequency of the input signal and the resonance frequency of the sensing element coincide. And when the output voltage signal is at its strongest, the frequency of the stimulus signal at the input electrode is the resonant frequency. Figure 1 depicts the structure of the sensing element. The proposed sensor uses PTZ (Piezoelectric Ceramic Transducer) and 3J53 alloy bonded with conductive adhesive as its sensing element.

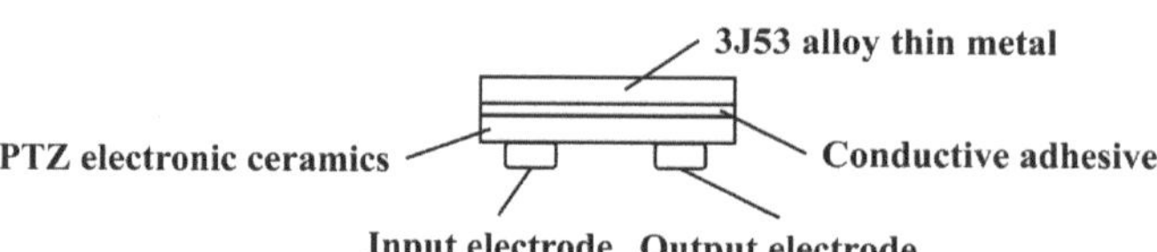

Figure 1. Ice sensing element structure.

The sensor measures the resonance frequency of the ice sensing element and the material on its surface to detect the presence of ice or water [28]. The sensor has a measuring range of 0~4 mm. The resonance frequency f of the electronic ceramic material is related to its equivalent stiffness and equivalent mass and is calculated as Equation (1). When ice or water are present on the surface of the sensing element, the ice or water form a new whole with the sensing element and change the original equivalent stiffness and equivalent mass of the sensing element. Different amounts of substances have different effects on the equivalent stiffness and equivalent mass of the sensing element, and a regular change in resonant frequency is produced, thus achieving the accurate detection of ice or water thickness.

$$f = \frac{1}{2\pi}\sqrt{\frac{K}{m}} \tag{1}$$

where *K* denotes the equivalent stiffness of the electronic ceramic material, *m* is the equivalent mass of the electronic ceramic material.

To monitor the ice thickness on a pavement surface using a sensor, the relationship between the sensor's resonant frequency and the water and ice film thickness needs to be clarified. In this study, different water and ice film thicknesses were prepared on the sensor element surface first. Then, the resonance frequency of the sensor with different film thicknesses of water and ice was recorded, as shown in Figure 2.

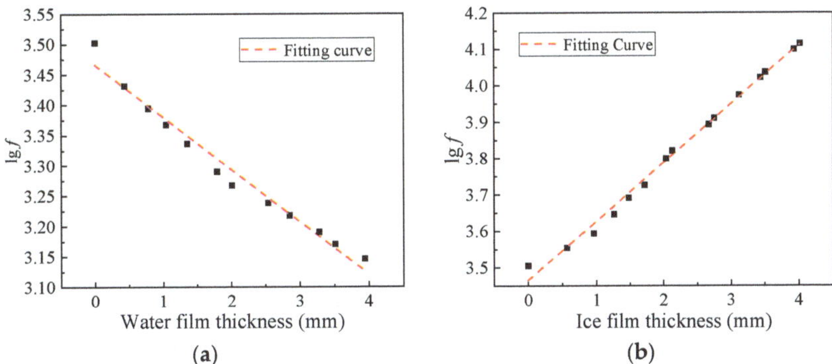

Figure 2. Fitting curve between resonant frequency and film thickness: (**a**) lg*f* and water film thickness; (**b**) lg*f* and ice film thickness.

Figure 2 shows that the initial lg*f* value of the sensor was 3.5 (when the sensor surface was air), which was the turning point to distinguish whether the material on the sensor surface was water or ice. Resonance frequency was linearly correlated with the amount of water or ice covering the sensor. The denary logarithm of resonance frequency *f* was less than 3.5 when water was present on the sensor surface. As the water film thickness increased, the resonance frequency decreased because when the water volume increased, the electronic ceramic material's equivalent stiffness was nearly unchanged, whereas the equivalent mass increased. When the sensor surface was covered with ice, the denary logarithm value of resonance frequency *f* exceeded 3.5. The resonant frequency increased with increasing ice thickness because the equivalent mass and stiffness of the electronic ceramic material increased, and the latter dominated the change in resonance frequency. According to this information, a sensor using piezoelectric technology can measure the thickness of ice or water to accurately acquire pavement surface information.

2.   Sensor Structure

The sensor in this study is composed of a shell, a base, an ice sensing element, a temperature sensor, a signal generator, and a microcomputer, as illustrated in Figure 3. The shell of the sensor was made of stainless steel to resist corrosion and loads, and the protection rank could reach IP68. For easy installation, a 74 mm diameter shell that was slightly smaller than the coring machine was used, and the height of the shell was 62 mm. By comprehensively considering the efficacy and size of the sensor, the AD9834 signal generator and STC12C5A60S2 microcontroller were selected to generate, collect, and transmit the electrical signals of the sensor. The signal generator and microcontroller was provided by Shenzhen Yatai Yingke Electronics Co., Ltd, Guangdong, China. Approximately 3000 RMB are required to produce a sensor with this configuration. This low-priced sensor is suitable for overlay applications in countries such as China, which has a vast network of roads.

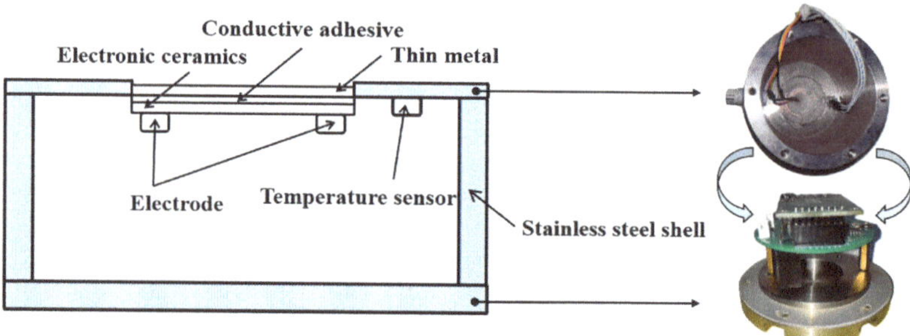

Figure 3. Schematic diagram of sensor structure.

2.1.2. Sensor Operating Characteristics

1. Sensor Calibration

The developed piezoelectric sensor is expected to maintain good working characteristics within the operation temperature range. Therefore, different ice thicknesses were prepared at different temperatures to calibrate the sensor and examine its measurement accuracy. The accuracy of the sensor was calculated as Equations (2) and (3). The output value of the sensor was automatically calculated by complying with the relationship in Figure 2. The actual thickness of the ice or water film was calculated as the ratio of its volume to the surface area of the sensor.

$$\varepsilon = \frac{|Thickness_A - Thickness_O|}{Thickness_A} \times 100\% \tag{2}$$

$$Thickness_A = \frac{Volume_{water/ice}}{Sensing\ element\ surface\ area} \tag{3}$$

where $Thickness_A$ denotes the actual thickness of ice or water film, mm; $Thickness_O$ is the output value of the sensor, mm.

As shown in Figure 4, all data points were concentrated near the standard line. The deviation between the output and actual values was small. The average maximum error of the output value was less than 0.08 mm, which was only 2% of the measurement range. The accuracy of the sensor at -10, -15, and -20 °C was 4%, 3.7%, and 3.5%, respectively. This result indicates that the sensor can work in harsh environments with good accuracy.

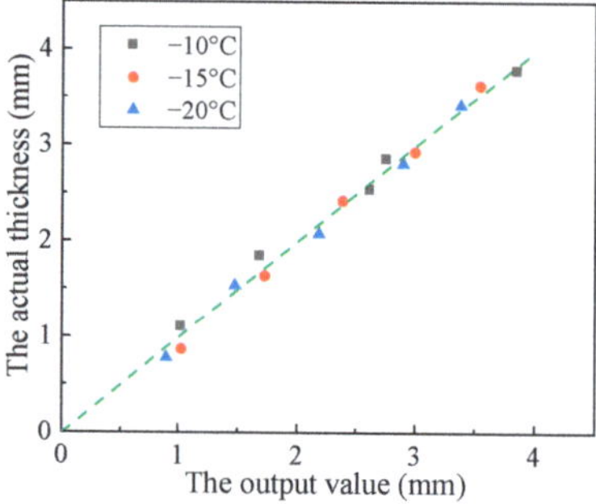

Figure 4. Accuracy of the sensor at different temperatures.

2. Sensor Installation

In this study, two sensors were installed on an expressway surface to verify the long-term service performance of the piezoelectric sensors. The installation site was located in Laiqu Expressway in Hebei Province, China. The installation process included the preparation of safety work, the installation of the sensors, and the establishment of data transmission pathways, as shown in Figure 5. During installation, the sensing element was kept flush with the pavement surface. Data transmission was achieved using a data transmission box. The data transmission box was mounted on the roadside, connected to the sensors, and powered by 220 V of AC power. The box could quickly transmit monitoring data to the host computer via a wireless network with a delay of less than 0.1 s, thus enabling real-time detection of pavement icing.

Figure 5. Installation of sensors in Laiqu Expressway in Hebei Province, China: (**a**) preparation of safety work; (**b**) grooving of pavement; (**c**) installation of the sensor; (**d**) rehabilitation of pavement and establishment of data transmission pathways.

After four months of operation, the sensors were consistently in good condition and had an average maximum error of less than 0.15 mm, as shown in Figure 6. The sensors consistently demonstrated good accuracy and reliability when subjected to vehicle loads, temperature loads, and harsh natural environments in these days. Abnormal data were obtained during the monitoring process due to the electromagnetic influence. The abnormal data had a short duration and a considerable difference from the normal value. Some data are shown in Figure 7. Only four abnormal data points out of 5500 data points were obtained, and the probability of an abnormal data appearance was less than 1‰. Thus, the abnormal data could be easily identified and did not affect the actual monitoring effect. In summary, the proposed piezoelectric sensor exhibited high precision and good durability in pavement surface status monitoring and could provide effective information collection support for pavement icing early warnings.

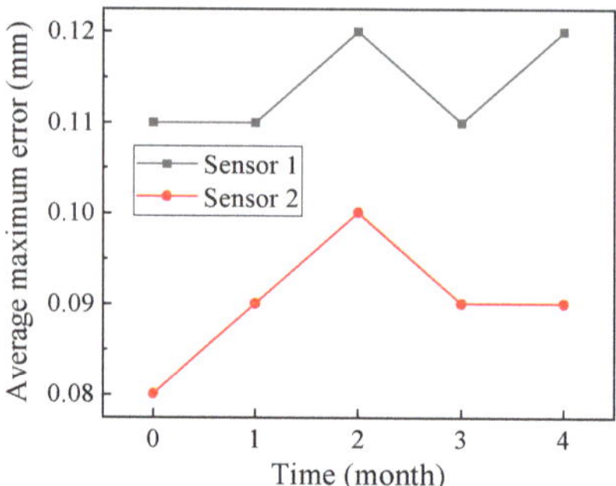

Figure 6. Average maximum error of the sensors in service.

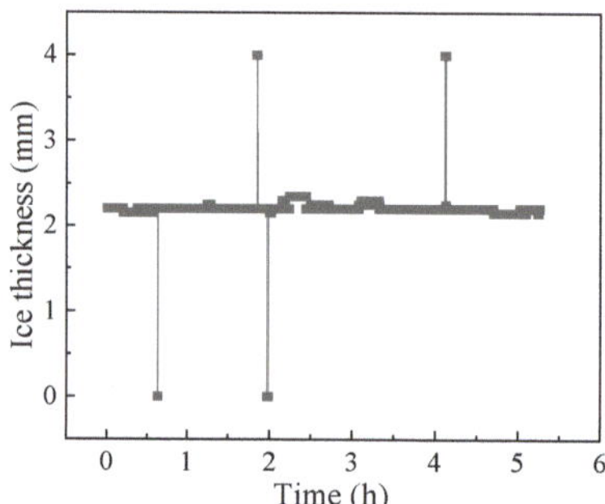

Figure 7. Examples of partial sensor monitoring data.

2.2. Method

The effects of different factors (pavement temperature, wind speed, and water depth) on ice formation on the pavement surface were investigated from the outside. Icing time is the time when the water is completely frozen and can be obtained through sensor sensing. The resonant frequency of the sensor changes continuously as the water freezes. The water is considered completely frozen when the resonant frequency no longer changes. Laboratory tests were performed to quantify the influence of various factors on icing time. Then, orthogonal design-based tests were conducted to determine the importance of the factors to icing time. Notably, the experimental data here can be used as a database for pavement icing early warning models.

The laboratory test equipment included a climatic chamber, a variable-speed fan, an ice sensor, and an AC-13 specimen whose length, width, and height were 300, 300, and 50 mm, respectively. The laboratory tests were performed in a climatic chamber with a length, width, and height of 3 m, as shown in Figure 8. The climatic chamber was assembled by the Harbin Institute of Technology Cold Region Road Research team. The environmental conditions included temperature, wind speed, and water depth. The temperature control accuracy of the climatic chamber was 0.1 °C, the wind speed was controlled by a variable-speed fan, and the water depth was prefabricated by quantitative weighing. The AC-13 asphalt mixture specimen was prepared using matrix 70 # asphalt and basalt aggregate. The parameters of the asphalt and aggregate met standard requirements. As shown in Figure 9, the sensor was embedded in the center of the specimen. A base was placed under the sensor and specimen to ensure that their surfaces were flush. The specimen and sensor were kept in the test environment for over an hour before testing, and water was placed at a temperature of 0 °C.

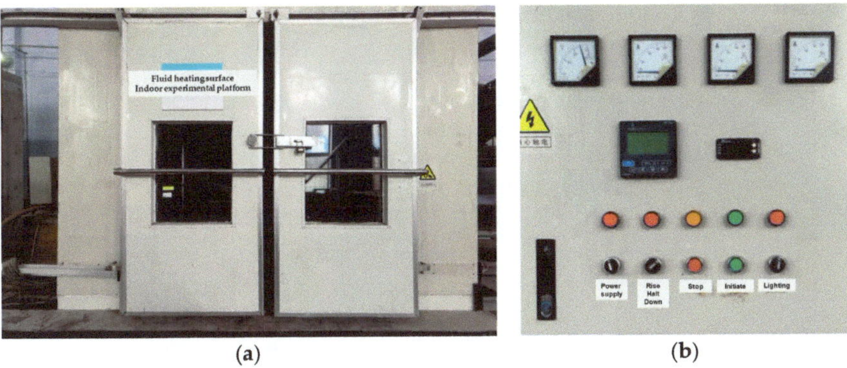

(a)   (b)

Figure 8. The climatic chamber: (**a**) chamber appearance; (**b**) temperature control system.

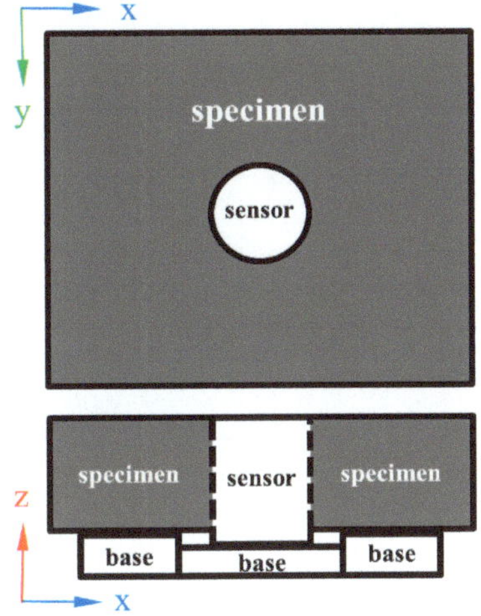

Figure 9. The sensor and the specimen.

The conditions of the experimental environment were determined based on the local climate. The climate data of Hebei Province obtained over the past three years show that the average daily minimum temperature in winter is about $-8\,^\circ$C, the average annual wind speed is less than 5 m/s, and the average precipitation is less than 4 mm. In consideration of these climatic conditions, the experimental temperatures were set to -2, -4, -6, and $-8\,^\circ$C. The wind speeds were set to 1, 2, 3, and 4 m/s, and the water depths were set to 1, 1.5, 2.5, and 3.5 mm. The equal-level orthogonal L16 (4 × 4) test design was applied to the four factors and levels, and a null column was added between the factors to estimate the random error. The three factors and the null column were classified as A, B, C, and D columns, respectively. The design scheme is given in Table 1.

Table 1. Factors and level of orthogonal test.

Level	Pavement Temperature (°C)	Wind Speed (m/s)	Water Depth (mm)	Null
1	−2	1	1	1
2	−4	2	1.5	2
3	−6	3	2.5	3
4	−8	4	3.5	4

3. Results and Discussion

Temperatures in Hebei Province exceed 0 °C during the day and are below 0 °C at night. Snowfall melts during the day and freezes to ice at night due to the temperature difference between day and night. The low-temperature conditions at night and the residual precipitation on the pavement surface are the key factors for pavement ice formation in Hebei Province. In addition, the presence of wind can accelerate the icing process. Therefore, pavement temperature, wind speed, and water depth were selected as experimental variables, and their specific effects on icing time were elucidated. The experimental data served as a basis for the nondestructive detection of pavement icing.

3.1. The Effect of Wind Speed on Pavement Icing

The presence of wind may accelerate energy dissipation and thus shorten the icing time. To analyze the effect of wind speed on icing time, icing time was tested at wind speeds of 1, 2, 3, and 4 m/s while maintaining a constant pavement temperature and water depth. Figure 10 depicts the relationship between icing time and wind speed. The color in the figure represents the icing time. As shown in the legend, the closer the color is to red, the longer the time is; the closer the color is to blue, the shorter the time is.

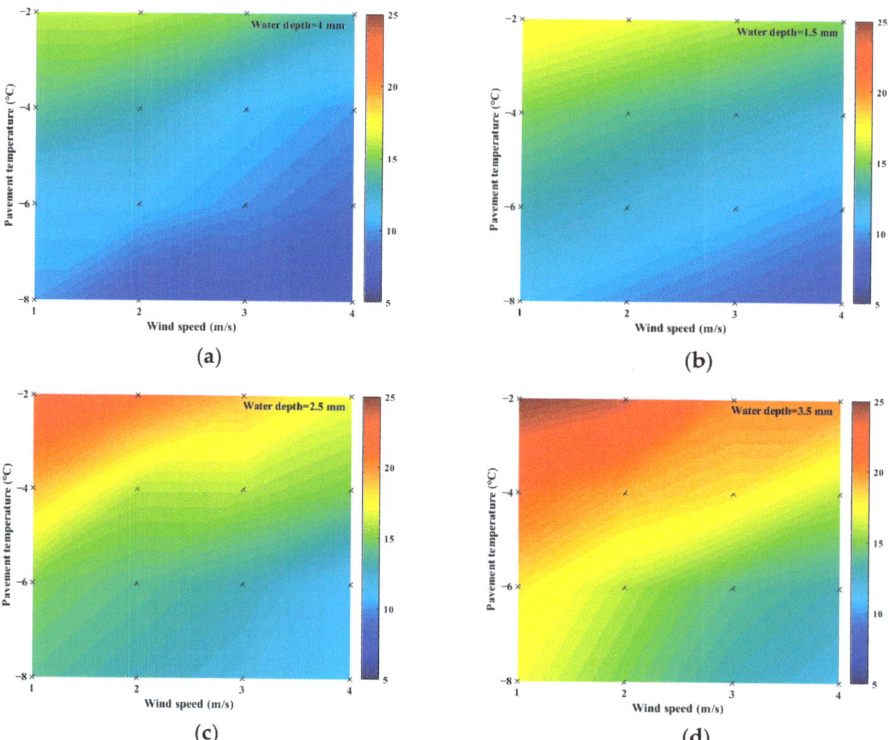

Figure 10. Effects of wind speed on icing time: (**a**) water depth = 1 mm; (**b**) water depth = 1.5 mm; (**c**) water depth = 2.5 mm; (**d**) water depth = 3.5 mm.

Under a certain pavement temperature and water depth, icing time increased with the decrease in wind speed. The shortening rate of icing time changed from 17% to 30% as the wind speed increased from 1 m/s to 4 m/s. This result proves that wind speed positively influenced icing time, but the degree of influence was small. The temperature of the water was close to that of the wind, and the internal energy of water could not easily exchange heat with the wind, causing a minor influence of wind speed on the icing time.

3.2. The Effect of Water Depth on Pavement Icing

In this section, it is assumed that the depths of the pavement surface structures are equal at all locations, and the amount of water on the pavement surface is quantified using the "water depth" (the value is equal to the water volume divided by the pavement area). To analyze the effect of water depth on the icing time, the experiment was conducted at water depths of 1, 1.5, 2.5, and 3.5 mm while keeping the temperature and wind speed constant. The results are presented in Figure 11.

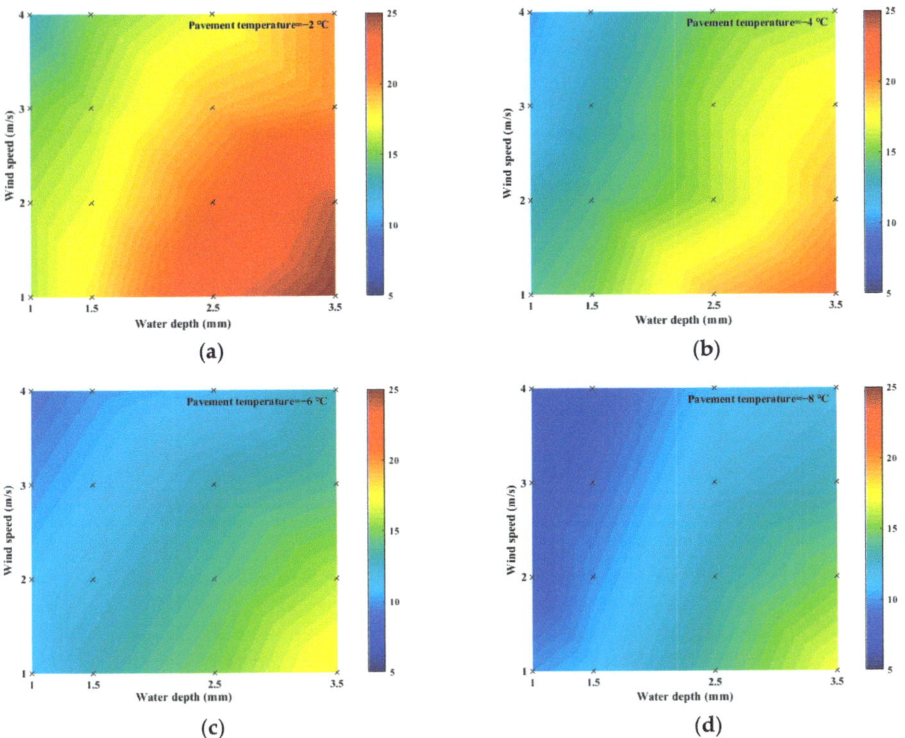

Figure 11. Effects of water depth on icing time: (**a**) pavement temperature = −2 °C; (**b**) pavement temperature = −4 °C; (**c**) pavement temperature = −6 °C; (**d**) pavement temperature = −8 °C.

When the pavement temperature and the wind speed remained unchanged, the icing time increased with the increase in the water depth. The shortening rate of the icing time changed from 25% to 47% as the water depth reduced from 3.5 mm to 1 mm. This result indicates that water depth negatively influenced icing time. Given the same contact area between water and other substances, the more water on the pavement surface, the slower the energy exchange between water and other substances, and the longer the icing time.

3.3. The Effect of Pavement Temperature on Pavement Icing

The pavement is in direct contact with water, and the pavement temperature will have some effect on the icing rate. To analyze the effect of pavement temperature on icing time, the experiment was conducted at pavement temperatures of −2, −4, −6, and −8 °C while keeping the water depth and wind speed constant. The results are presented in Figure 12.

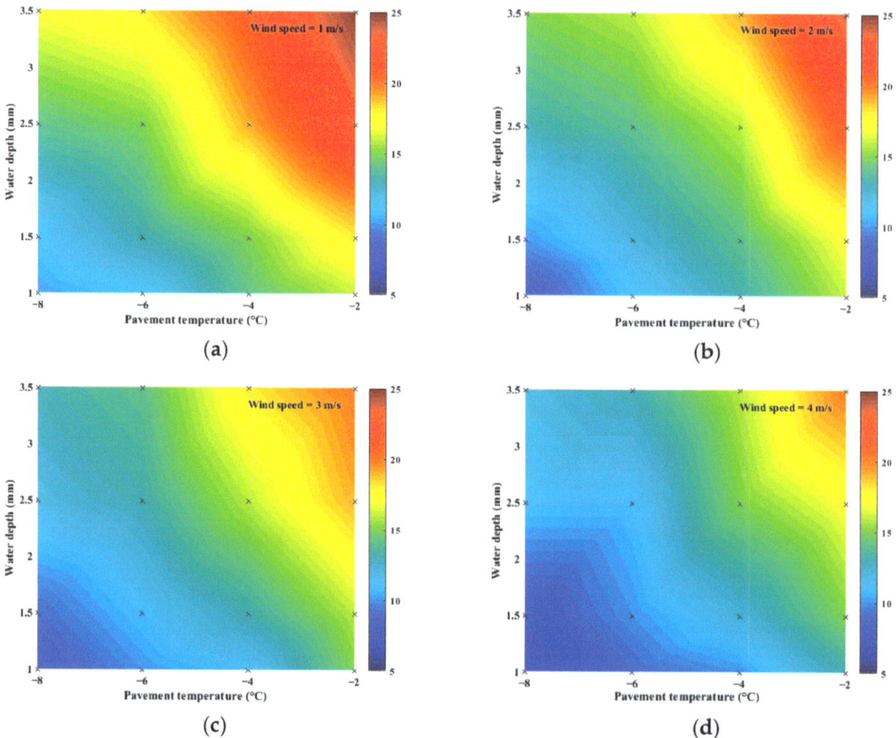

Figure 12. Effects of pavement temperature on icing time: (**a**) wind speed = 1 m/s; (**b**) wind speed = 2 m/s; (**c**) wind speed = 3 m/s; (**d**) wind speed = 4 m/s.

For a given wind speed and water depth, the icing time increased with the decrease in the pavement temperature. The shortening rate of icing time varied from 32% to 50% and the pavement temperature decreased from –2 °C to –8 °C. The result indicates that the pavement temperature positively influenced the icing time. The effect of the pavement temperature on icing time is greater than the wind speed and water depth, which is identical to the conclusion of the orthogonal test above. The pavement temperature was lower than that of water, which made it convenient for water to release heat and freeze. Lower pavement temperatures can accelerate the energy exchange between water and pavement, resulting in faster condensation nucleation and crystallization of water.

In general, the essence of pavement surface icing is that water loses heat, resulting in a decrease in water molecular kinetic energy, condensation, nucleation, and crystallization. According to the above test results, the icing time increases with the increase in pavement temperature, the slowing of wind speed,* and the thickening of water depth. The maximum icing time is more than 25 min under the conditions of −2 °C pavement temperature, 1 m/s wind speed, and 3.5 mm water depth. And the minimum icing time is less than 8 min when the temperature is −8 °C, the wind speed is 4 m/s, and the water depth is 3.5 mm.

3.4. Significance Analysis of Variables

An equal-level orthogonal L16 (4 × 4) test was performed to compare the significance of the effects of pavement temperature, wind speed, and water depth on pavement icing. The variables were statistically ranked in terms of the degree of influence they had on icing time. The design and results of the experiment are presented in Table 2.

Table 2. Factors and level of orthogonal test.

Number	1	2	3	4	5	6	7	8	9	10	11	12	13	14	15	16	K1	K2	K3	K4	Range
Pavement temperature (°C)	−2	−6	−4	−2	−6	−6	−4	−2	−8	−8	−4	−2	−8	−6	−8	−4	73	62	50	44	29
Wind speed (m/s)	4	2	1	2	4	3	3	1	4	3	2	3	1	1	2	4	64	61	55	49	15
Water depth (mm)	2.5	2.5	3.5	3.5	3.5	1	2.5	1	1	3.5	1	1.5	2.5	1.5	1.5	1.5	46	51	61	71	25
Null	3	1	3	4	2	3	4	1	4	1	2	2	2	4	3	1	55	56	58	60	5
Icing time (min)	17	14	21	24	13	10	16	16	7	13	13	16	14	13	10	12	-	-	-	-	-

An orthogonal test usually uses ranges to distinguish primary and secondary factors. As indicated by the range analysis in Table 2, the ranges of the pavement temperature, wind speed, water depth, and the null column reached 29, 15, 25, and 5. The pavement temperature was the main factor that affected the icing time of pavements, followed by water depth and wind speed. In addition, Table 2 shows that the three factors with the highest levels were K1, K1, and K4, and A1B1C4 was the longest test condition of icing time. Subsequently, to analyze the significance of factors, the variance of the test results was calculated to conduct an F-test, as shown in Table 3.

Table 3. The variance analysis of orthogonal test.

Factor	Degree of Freedom	MS	F	p
Pavement temperature	3	41.562	33.814	0.008
Wind speed	3	11.062	9	0.052
Water depth	3	30.729	25	0.013
Error	3	1.229	-	-

According to Table 3, the F values of pavement temperature, wind speed, and water depth were 33.814, 9, and 25, respectively. All F values were greater than the F critical value ($F(3,3,0.9) = 5.36$), indicating that the three factors significantly influenced icing time. In addition, the smaller the p value was, the more significant the result was. The p values of three factors were small, indicating that these factors significantly influenced icing time.

Overall, the pavement temperature, wind speed, and water depth had significant effects on icing time, with the effects of pavement temperature and wind speed being positive and the effects of water depth being negative. Pavement temperature had the greatest effect on icing time (up to 50%), followed by water depth; the effect of wind speed was the smallest (as small as 17%). However, the effects of the three factors on icing time were nonlinear, and the coupling between them was difficult to explain by a unified empirical formula.

4. Early Warning Model

The future of road engineering is nondestructive road facility identification and warning, which is made possible by quick statistical analysis and potent machine learning techniques [29–31]. Accurate road icing warnings based on meteorological data are required to achieve nondestructive detection of road ice. Given that the factors contributing to pavement icing are interrelated and complex, pavement icing early warning should comprehensively consider the nonlinear relationship of multiple variables. An artificial neural network (ANN) is an adaptive multilayer network that is particularly suitable for solving this problem due to its complex internal mechanisms, which have been proven by mathematical theory.

The single hidden-layer backpropagation (BP) training algorithm of ANN was selected to predict the pavement icing time in this study, as shown in Figure 13.

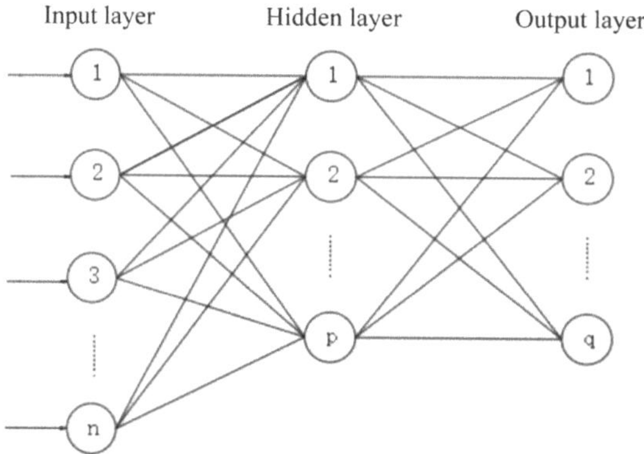

Figure 13. The structure of single hidden-layer neural network.

S-shaped neurons were used in the hidden layer, and linear neurons were used in the output layer. The transfer function is a (0,1) S-shaped function, which is shown in Equation (4).

$$f(x) = \frac{1}{1 + e^{-x}} \tag{4}$$

The operation of the network includes a series of steps: network initialization, importing sample data, the output calculation of the network, the partial derivatives calculation of the error function, adaptation of the connection weight, and the calculation of the sample error. The final sample error formula is shown in Equation (5). The process is terminated if the error satisfies the predetermined accuracy standards or has been trained a maximum number of times; otherwise, network training is repeated.

$$E = \frac{1}{2m} \sum_{k=1}^{m} \sum_{s=1}^{q} (t_s(k) - yo_s(k))^2 \tag{5}$$

where m denotes the number of samples; $t(k)$ is the desired output; $yo(k)$ is the output of network.

The results in Section 4 show that pavement temperature, water depth, and wind speed had strong correlations with icing time. Thus, the three meteorological parameters were adopted as input parameters, and icing time was employed as an output parameter. In addition, the number of hidden units and the learning rate, which have a significant effect on calculation speed and result accuracy, served as vital parameters of the BP neural network. The prediction accuracy of the model with different numbers of hidden units and learning rates was determined to reduce the prediction error. The database consisted of 64 groups of test data from the previous laboratory experiment, as shown in Table 4. The experimental temperatures were set to −2, −4, −6, and −8 °C. The wind speeds were set to 1, 2, 3, and 4 m/s, and the water depths were set to 1, 1.5, 2.5, and 3.5 mm. Icing times were recorded for 64 combinations of the above conditions. The experimental data were randomly arranged to ensure the randomness of the model input data. Fifty-one groups (80%) of randomly selected laboratory test data were used to train the prediction model, and the remaining test data (20% of randomly selected laboratory test data) served as a testing set to test the prediction accuracy of the model.

Table 4. Data basis of the artificial neural network.

No.	Pavement Temperature (°C)	Wind Speed (m/s)	Water Depth (mm)	Icing Time (min)	No.	Pavement Temperature (°C)	Wind Speed (m/s)	Water Depth (mm)	Icing Time (min)
1	−6	4	1	9	33	−6	2	1.5	12
2	−4	2	3.5	19	34	−2	4	1	13
3	−6	3	3.5	14	35	−2	4	1.5	15
4	−6	1	1	11	36	−4	4	3.5	16
5	−6	3	1	10	37	−8	2	2.5	13
6	−8	4	3.5	12	38	−8	4	2.5	11
7	−8	3	1	8	39	−2	1	2.5	23
8	−2	3	2.5	19	40	−6	3	2.5	13
9	−6	1	1.5	13	41	−4	4	1	10
10	−2	4	3.5	20	42	−4	2	2.5	16
11	−8	1	3.5	17	43	−6	4	2.5	11
12	−2	2	1	16	44	−6	1	3.5	18
13	−6	3	1.5	11	45	−2	1	1.5	18
14	−8	1	2.5	14	46	−4	4	2.5	15
15	−8	4	1.5	8	47	−6	2	2.5	14
16	−8	2	1.5	10	48	−8	2	1	8
17	−4	1	2.5	19	49	−8	3	3.5	13
18	−8	3	1.5	9	50	−2	3	1.5	16
19	−6	2	3.5	16	51	−2	1	3.5	25
20	−8	1	1	10	52	−4	3	2.5	16
21	−4	3	1	11	53	−4	4	1.5	12
22	−2	1	1	16	54	−6	1	2.5	15
23	−8	4	1	7	55	−4	3	3.5	18
24	−8	2	3.5	15	56	−8	1	1.5	11
25	−4	1	3.5	21	57	−6	4	3.5	13
26	−4	1	1	14	58	−2	3	3.5	20
27	−2	2	1.5	17	59	−2	4	2.5	17
28	−2	2	3.5	24	60	−4	2	1.5	14

4.1. Number of Hidden Units

A hidden unit is a fundamental computing unit in a neural network's hidden layer that receives and processes data and corrects the weight coefficients. The hidden layer of a neural network contains several hidden units, and the number of hidden units can be used to determine the goodness of fit of the neural network. A model with a small number of hidden units cannot easily acquire sufficient information from the learning set, which may cause model underfitting. Meanwhile, a training network with too many hidden units may cause model overfitting and insufficient generalization ability. In this study, the numbers of hidden units were 3, 5, 7, and 9 when the BP neural network prediction model was implemented on the training set. The prediction accuracy of the model was analyzed based on the testing set. The results are shown in Figure 14 and Table 5.

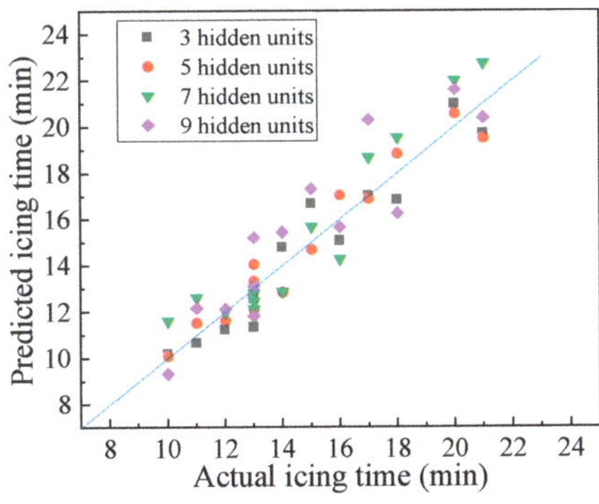

Figure 14. Deviation of predicted values of models with different number of neurons.

Table 5. Model prediction under different number of neurons.

Number of Hidden Units	Average Error (min)	Maximum Error (min)	Accuracy (%)	Root Mean Square Error (min)	Pearson Correlation Coefficient
3	0.81	1.69	90.6	0.96	0.961
5	0.67	1.50	91.7	0.79	0.970
7	1.18	2.00	88.9	1.33	0.953
9	1.30	3.31	81.6	1.59	0.913

According to Figure 14 and Table 5, accuracy initially increased, then decreased with the increase in the number of hidden units. When the number of hidden units was 5, the average, maximum, and root mean square errors of the model were 0.67, 1.50, and 0.79 min, respectively, which are the minimum values for different models. This result indicates that the predicted value of the current model is the closest to the real value. The Pearson test showed that the model's goodness of fit was the highest. Thus, the optimal number of hidden units for the single-layer neural network model was determined to be 5. The neural network with five hidden units can predict pavement icing time efficiently.

4.2. Learning Rate

BP neural network models are typically trained with the gradient descent method, and the learning rate is related to the distance that determines how far the weights move in the gradient direction. The learning rate determines whether and when the objective function converges. Low learning rates decrease the convergence speed of the model. However, if the learning rate is too high, network shock or even non-convergence may occur. The BP neural network prediction model with learning rates of 0.5, 0.6, 0.7, and 0.8 was trained in this work. Then, the prediction accuracy of the model was analyzed based on the testing set. The results are shown in Figure 15 and Table 6.

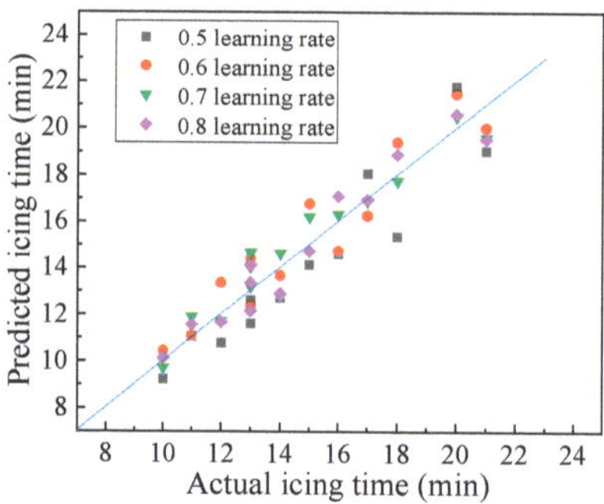

Figure 15. Deviation of predicted values of models with different learning rates.

Table 6. Model prediction under different learning rates.

Learning Rate	Average Error (min)	Maximum Error (min)	Accuracy (%)	Root Mean Square Error (min)	Pearson Correlation Coefficient
0.5	1.19	2.64	85.3	1.36	0.921
0.6	0.96	1.73	90.4	1.08	0.943
0.7	0.67	1.50	91.7	0.79	0.970
0.8	0.67	1.64	90.9	0.83	0.963

Figure 15 and Table 6 show that with the increase in the learning rate, accuracy initially increased then decreased, and this pattern was consistent with the change law of the number of hidden units. When the learning rate was 0.7, the average, maximum, and root mean square errors of the model were the smallest, and the Pearson correlation coefficient was the largest. The optimal learning rate of the neural network model was determined to be 0.7. Therefore, the neural network with five hidden units and a learning rate of 0.7 predicts the pavement icing time the best, and it may be valuable for pavement icing warning work.

4.3. The Validation of The Model

The accuracy and validity of the model were assessed using the validation dataset. The three meteorological conditions, namely, temperature, wind speed, and water depth, were inputted into the BP neural network model to obtain the predicted icing time. The prediction results of the model are shown in Figure 16 and Table 7.

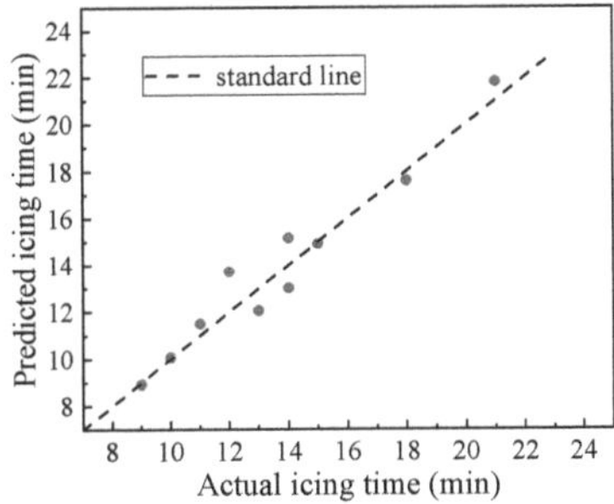

Figure 16. Comparison between the actual icing time and the icing time predicted by B.P. neural network model.

Table 7. Prediction results of the B.P. neural network model.

Average Error (min)	Maximum Error (min)	Accuracy (%)	Root Mean Square Error (min)	Pearson Correlation Coefficient
0.71	1.68	90.7	0.83	0.986

As indicated in Figure 16 and Table 7, the predicted values of the icing time obtained from the BP neural network model had small errors, with the mean, maximum, and root mean square errors being 0.71, 1.83, and 0.83 min, respectively. The Pearson correlation coefficient between the predicted and measured values was 0.986, indicating a good correlation between the two. In addition, the model's prediction accuracy reached 90.7, which indicates an accurate prediction of the pavement icing time.

To sum up, the single-layer BP neural network model that uses pavement temperature, water depth, and wind speed as the input parameters successfully predicted the pavement icing time. The model with a learning rate of 0.7 and five hidden units had the best prediction effect on pavement icing, and its accuracy reached 91.7%. The model can be used for pavement pre-icing warning and can help prevent traffic accidents. Given that the BP neural network model is an adaptive learning network, the parameters of this model are applicable to areas at the same latitude and longitude as Hebei Province.

5. Conclusions

In this study, an icing nondestructive detection and early warning system for asphalt pavements was established. First, the principle and characteristics of the sensors were introduced, and the performance of the sensors was investigated. Second, the icing time under various operating situations was studied, and a highly accurate, self-adaptive early warning model for pavement icing was developed. The main research results are as follows:

- The piezoelectric sensors were designed to quantitatively indicate the substance (i.e., air, water, or ice) covering the sensors' surfaces in accordance with the function of substance thickness and resonance frequency. lgf showed a linear relationship with ice and water thickness; it decreased with the increase in water thickness and increased with the increase in ice thickness. The piezoelectric sensors exhibited the advantages of low cost, strong anti-interference ability, high measurement accuracy, and long survival time and are thus suitable for pavement surface condition monitoring.
- In the experiments, the effects of pavement temperature, wind speed, and water depth on pavement icing time were studied. The results showed that icing time increased with the increase in pavement temperature, the slowing of wind speed, and the thickening of water depth. Pavement temperature was the critical factor in determining icing time, and it could reduce icing time by 50% under normal winter conditions in Hebei Province. With the action of the three factors, the longest time for pavement icing exceeded 25 min, and the shortest time was only 8 min.
- On the basis of the influence law of wind speed, water depth, and pavement temperature, a warning for pavement icing was provided by the BP neural network model. The BP neural network model with a learning rate of 0.7 and five hidden units exhibited optimal prediction performance in pavement icing early warning, and its prediction accuracy was as high as 91.7%.

In this study, the early warning system was adopted to warn of ice caused by freeze–thaw of precipitation and black ice attributed to air humidity in central and southern China, without considering the water type (contamination) and the snowfall in northern areas. The influence of other contaminations, such as sand and lubricating oil, on the detection effects was also not taken into account. In addition, the usefulness of the early warning system needs to be further improved, and more powerful statistical and machine learning methods should be used. The focus of the follow-up studies will be centered on the above aspects, depending on the climatic conditions in cold regions. Thus, the general applicability of the pavement icing early warning system can be improved to better support the safe operation of expressway traffic.

Author Contributions: Conceptualization, Y.T. and J.L. (Jilu Li); methodology, H.X. and J.L. (Jilu Li); software, J.L. (Jilu Li) and H.M.; validation, B.Z.; formal analysis, J.L. (Jilu Li); investigation, Y.T., H.X. and J.L. (Jilu Li); resources, H.M., W.S. and J.L. (Jie Liu); data curation, W.S.; writing—original draft preparation, J.L. (Jilu Li); writing—review and editing, Y.T. and H.X.; visualization, W.S. and B.Z.; supervision, J.L. (Jie Liu); project administration, H.M.; funding acquisition, Y.T. and H.X. All authors have read and agreed to the published version of the manuscript.

Funding: This research was funded by the National Natural Science Foundation of China Joint Fund for Regional Innovation and Development, grant number No. U20A20315, Key research project of Heilongjiang Province, grant number No. 2022ZXJ02A02, Key R&D Plan Program of Hebei Province, grant number No. 20375405D, and Science and technology project of Qinghai Province, grant number No. 2021-QY-207.

Institutional Review Board Statement: Not applicable.

Informed Consent Statement: Not applicable.

Data Availability Statement: Not applicable.

Acknowledgments: Not applicable.

Conflicts of Interest: The authors declare no conflict of interest.

References

1. Tan, T.; Xing, C.; Tan, Y.; Gong, X. Safety Aspects on Icy Asphalt Pavement in Cold Region through Field Investigations. *Cold Reg. Sci. Technol.* **2019**, *161*, 21–31. [CrossRef]
2. Chen, F.; Chen, S. Reliability-Based Assessment of Vehicle Safety in Adverse Driving Conditions. *Transp. Res. Part C Emerg. Technol.* **2011**, *19*, 156–168. [CrossRef]

3. Habibi Nokhandan, M.; Bazrafshan, J.; Ghorbani, K. A Quantitative Analysis of Risk Based on Climatic Factors on the Roads in Iran. *Meteorol. Appl.* **2008**, *15*, 347–357. [CrossRef]

4. Veneziano, D.; Ye, Z.; Turnbull, I. Speed Impacts of an Icy Curve Warning System. *IET Intell. Transp. Syst.* **2014**, *8*, 93–101. [CrossRef]

5. Ma, X.; Ruan, C. Method for Black Ice Detection on Roads Using Tri-Wavelength Backscattering Measurements. *Appl. Opt.* **2020**, *59*, 7242–7246. [CrossRef]

6. Zhang, J.; Wang, P.; Wang, H.; Soares, L.; Pei, J.; Ge, J. Development and Verification of Integrated Photoelectric System for Noncontact Detection of Pavement Ponding and Freezing. *Struct. Control Health Monit.* **2021**, *28*, e2719. [CrossRef]

7. Kim, J.; Kim, E.; Kim, D. A Black Ice Detection Method Based on 1-Dimensional CNN Using mmWave Sensor Backscattering. *Remote Sens.* **2022**, *14*, 5252. [CrossRef]

8. Shah, A.; Niksan, O.; Zarifi, M.H. Planar Microwave Sensor for Localized Ice and Snow Sensing. In Proceedings of the International Conference on Icing of Aircraft, Engines, and Structures, Vienna, Austria, 20–22 June 2023; SAE International: Warrendale, PA, USA, 2023.

9. Du, Y.; Jiang, C.; Xiao, H.; Li, H. Study of SAW Ice Sensor Application to Highway Airstrip. In Proceedings of the 2009 Symposium on Piezoelectricity, Acoustic Waves, and Device Applications (SPAWDA 2009), Wuhan, China,, 17–20 December 2009; p. 29.

10. Liu, T.; Pan, Q.; Sanchez, J.; Sun, S.; Wang, N.; Yu, H. Prototype Decision Support System for Black Ice Detection and Road Closure Control. *IEEE Intell. Transp. Syst. Mag.* **2017**, *9*, 91–102. [CrossRef]

11. Troiano, A.; Pasero, E.; Mesin, L. New System for Detecting Road Ice Formation. *IEEE Trans. Instrum. Meas.* **2011**, *60*, 1091–1101. [CrossRef]

12. Moon, N.-W.; Choi, J.-H.; Kim, Y.-H. Polarimetric Monitoring of Road Surface Status for Safety Driving of Vehicles. In Proceedings of the 2012 9th European Radar Conference, Amsterdam, The Netherlands, 31 October–2 November 2012; pp. 10–13.

13. Ikiades, A.A. Direct Ice Detection Based on Fiber Optic Sensor Architecture. *Appl. Phys. Lett.* **2007**, *91*, 104104. [CrossRef]

14. Riehm, M.; Gustavsson, T.; Bogren, J.; Jansson, P.-E. Ice Formation Detection on Road Surfaces Using Infrared Thermometry. *Cold Reg. Sci. Technol.* **2012**, *83–84*, 71–76. [CrossRef]

15. Zhang, C.; Xiao, C.; Li, S.; Guo, X.; Wang, Q.; He, Y.; Lv, H.; Yan, H.; Liu, D. A Novel Fiber-Optic Ice Sensor to Identify Ice Types Based on Total Reflection. *Sensors* **2023**, *23*, 3996. [CrossRef] [PubMed]

16. Zhang, Z.; Zhang, D.; Jiang, X.; Huang, H.; Gao, D.W. Study of the Icing Growth Characteristic and Its Influencing Factors for Different Types of Insulators. *Turk. J. Electr. Eng. Comput. Sci.* **2016**, *24*, 1571–1585. [CrossRef]

17. Xu, H.; Zheng, J.; Li, P.; Wang, Q. Road Icing Forecasting and Detecting System. *AIP Conf. Proc.* **2017**, *1839*, 020089. [CrossRef]

18. Samodurova, T. Estimation of Significance the Parameters, Influencing on Road Ice Formation (the Results of Computing Experiment). In Proceedings of the 11th International Road Weather Conference, Sapporo, Japan, 26–28 January 2002; Available online: https://trid.trb.org/view/644932 (accessed on 29 September 2023).

19. Samodurova, T. Physicostatistical models for road ice formation forecast. *Sci. Her. Voronezh State Univ. Archit. Civ. Eng.* **2009**, *2*, 36–42.

20. Lee, N.S.T.; Karimi, H.A.; Krakiwsky, E.J. Road Information Systems; Impact of Geographic Information Systems Technology to Automatic Vehicle Navigation and Guidance. In Proceedings of the Conference Record of papers presented at the First Vehicle Navigation and Information Systems Conference (VNIS '89), Toronto, ON, Canada, 11–13 September 1989; pp. 347–352.

21. Perrin, J.; Hansen, B. Road Weather Information Systems: Finding out What People Want to Know. In Proceedings of the World Multi-conference on Systemics, Cybernetics and Informatics, Orlando, FL, USA, 14–18 July 2002.

22. Al-Harbi, M.; Yassin, M.F.; Bin Shams, M. Stochastic Modeling of the Impact of Meteorological Conditions on Road Traffic Accidents. *Stoch. Environ. Res. Risk Assess.* **2012**, *26*, 739–750. [CrossRef]

23. Ryerson, C.C.; Ramsay, A.C. Quantitative Ice Accretion Information from the Automated Surface Observing System. *J. Appl. Meteorol. Climatol.* **2007**, *46*, 1423–1437. [CrossRef]

24. Korotenko, K. An Automated System for Prediction of Icing on the Road. In Proceedings of the Computational Science—ICCS 2002, Amsterdam, The Netherlands, 21–24 April 2002; Sloot, P.M.A., Hoekstra, A.G., Tan, C.J.K., Dongarra, J.J., Eds.; Springer: Berlin/Heidelberg, Germany, 2002; pp. 1193–1200.

25. Martorina, S.; Nicola, L. Monitoring and Forecast Service for Ice Formation over Mountain Road Surface. *Hrvatski Meteorološki Časopis* **2005**, *40*, 414–417.

26. Mahura, A.; Petersen, C.; Sass, B. *Road Icing Conditions in Denmark*; DMI Sci. Report 08-03; DMI Ministry of Climate and Energy: Copenhagen, Denmark, 2008; 25p, ISBN 9788774785675.

27. Ye, Z.; Veneziano, D.; Turnbull, I. Safety Effects of Icy-Curve Warning Systems. *Transp. Res. Rec.* **2012**, *2318*, 83–89. [CrossRef]

28. Roy, S.; Izad, A.; DeAnna, R.G.; Mehregany, M. Smart Ice Detection Systems Based on Resonant Piezoelectric Transducers. *Sens. Actuators A Phys.* **1998**, *69*, 243–250. [CrossRef]

29. Lin, Z.; Wang, H.; Li, S. Pavement Anomaly Detection Based on Transformer and Self-Supervised Learning. *Autom. Constr.* **2022**, *143*, 104544. [CrossRef]

30. Luo, Z.; Wang, H.; Li, S. Prediction of International Roughness Index Based on Stacking Fusion Model. *Sustainability* **2022**, *14*, 6949. [CrossRef]

31. Wang, H.; Zhang, X.; Jiang, S. A Laboratory and Field Universal Estimation Method for Tire–Pavement Interaction Noise (TPIN) Based on 3D Image Technology. *Sustainability* **2022**, *14*, 12066. [CrossRef]

materials

Article

Investigation of the Relationship between Permanent Deformation and Dynamic Modulus Performance for Bearing-Layer Asphalt Mixture

Weidong Ji [1], Yunrui Meng [1], Yunlong Shang [2], Xiwei Zhou [1] and Huining Xu [1,*]

1 School of Transportation Science and Engineering, Harbin Institute of Technology, Harbin 150001, China; wedoji@163.com (W.J.); yunruimeng2@gmail.com (Y.M.); 20b932029@stu.hit.edu.cn (X.Z.)
2 Heilongjiang Transportation Investment Group Co., Harbin 150028, China; myrstu1@163.com
* Correspondence: xuhn@hit.edu.cn; Tel.: +86-0451-862-82120

Abstract: Of major concern is the lack of correlation between the material design and structural function of asphalt pavement in China. The objective of this paper is to identify the layer in asphalt pavement where permanent deformation occurs most seriously and to propose a control index for that layer's asphalt mixture. The permanent deformation of each layer was determined through the utilization of thickness measurements obtained from field cores. The results indicate that the reduction in thickness is more significant in the driving lane than in the ridge band and shoulder. This phenomenon can be attributed to the intensified densification and shearing deformation that arise from the combined impacts of recurrent axle loads and high temperatures. Compared to surface and base layers, the bearing layer is the primary area of concern for permanent deformation in asphalt pavement. Therefore, it is imperative to incorporate the ability of bearing-layer asphalt mixture to withstand permanent deformation as a crucial design parameter. The dynamic modulus of the bearing-layer asphalt mixture is significantly influenced by the type of asphalt, gradation, and asphalt content, compared to other design parameters. Based on the relationship established between dynamic modulus and dynamic stability, with creep rate as the intermediate term, a control standard was proposed to evaluate the permanent deformation of the bearing-layer asphalt mixture. This study can provide reasonable and effective guidance for prolonging pavement life and improving pavement performance.

Keywords: asphalt mixture; bearing layer; permanent deformation; dynamic modulus; control standard

Citation: Ji, W.; Meng, Y.; Shang, Y.; Zhou, X.; Xu, H. Investigation of the Relationship between Permanent Deformation and Dynamic Modulus Performance for Bearing-Layer Asphalt Mixture. *Materials* **2023**, *16*, 6409. https://doi.org/10.3390/ma16196409

Academic Editor: Giovanni Polacco

Received: 31 August 2023
Revised: 20 September 2023
Accepted: 21 September 2023
Published: 26 September 2023

1. Introduction

Asphalt pavement has become the most important form of road paving in China due to its good driving comfort, excellent road performance, and convenient maintenance characteristics [1,2]. During the service life of asphalt pavement, various types of distress commonly occur. Therefore, understanding the material parameters and pavement conditions is crucial for durability and performance assessments. The composition, layer thickness, and material properties (e.g., density, stiffness, elasticity, and thermal conductivity) of pavement material are parameters that need to be taken into consideration [3]. To determine material parameters, certain algorithms have been created which are based on the mechanical features of pavement. For example, the multi-level inverse algorithm is designed to simultaneously identify the stiffness and thickness of pavement models [4]. Utilizing visual and non-destructive testing methods, recent research has advanced surface quality assessment [5,6]. Garbowski et al. [7] used three-dimensional laser scanning of a road to identify pavement deterioration type and its quantity. De Blasiis et al. [8] developed a program for processing 3D point cloud data to identify and quantify a few types of pavement distress. By utilizing these technologies, we can assess and study the existing pavement to direct the pavement design.

Designing the asphalt mixtures is a key factor in ensuring the quality of asphalt pavement. Asphalt pavement is constructed layer by layer, and each layer has its own functional emphasis and stress characteristics [9]. Specifically, for a pavement structure with three layers of asphalt pavement, the surface layer ensures driving comfort, the bearing layer mainly resists permanent deformation, and the base layer is more critical in resisting fatigue [10,11]. With the increase in pavement service life and traffic loading, permanent deformation has become one of the main sources of the distress of asphalt pavement. Permanent deformation has a significant impact on the performance of asphalt pavement, particularly as the permanent deformations trap water and cause hydroplaning [12,13]. Moreover, significant permanent deformation can lead to major structural failures [14]. Hence, the permanent deformation not only reduces the service life, but it may also affect the safety of highway users.

It is well known that the resistance to permanent deformation is closely related to the pavement materials and structures. According to the research conducted by Xu et al. [10], the primary source of vertical deformation in asphalt pavement is the shearing deformation of its bearing layer. From the perspective of the mechanical response of asphalt pavement under traffic load, the bearing layer is in the high-shear-stress area. Li et al. [11] discovered that the bearing layer of the asphalt mixture became the primary compressed layer as permanent deformation formed, based on their measurements of the thickness of field cores. The study by Zhao et al. [15] delved into the factors that impact the permanent deformation of asphalt pavement. The researchers analyzed 180 field cores and determined that road age and equivalent single-axle loads were the primary determinants of permanent deformation of the pavement structure. Through the implementation of the Hamburg wheel tracking test, Zhu et al. [16] appraised the permanent deformation of the bearing layer of an existing highway. Evidence from the literature suggests that the bearing layer of asphalt can suffer permanent deformation, and this should be taken into account when designing asphalt mixtures. At present, the Marshall method is used in China for asphalt mixture design. The volume parameter is typically adopted as the primary design index, which utilizes the dynamic stability as the verification parameter for permanent deformation. It creates a disconnect between the material and structural design of asphalt pavement and makes a significant difference in the anti-permanent-deformation requirements of the bearing-layer asphalt mixture. Therefore, it is necessary to study the design index for bearing-layer asphalt mixture based on permanent deformation.

Various endeavors have been undertaken to establish testing methodologies, and evaluation parameters that can effectively measure the permanent deformation of asphalt mixtures, including the dynamic modulus test, dynamic creep test, repeated-load permanent deformation test, Hamburg wheel tracking test, etc. [17–20]. The relevant research results of the Long-Term Pavement Research Program in the United States show that the dynamic modulus has a good correlation with the rut depth in the field, which can be used to evaluate the permanent deformation of asphalt mixture [21]. The permanent deformation of six asphalt mixtures were investigated by Zhang et al. [22], who discovered that the dynamic creep is strongly correlated with the outcomes of dynamic modulus and repeated-load permanent deformation tests. Chaturabong et al. [23] observed that dry Hamburg wheel tracking was strongly correlated with the deformation in the second stage of the dynamic creep test, and that it can simulate field behavior very well. Witczak et al. conducted a comprehensive study on the dynamic modulus. The test results show that there is a high correlation between permanent deformation and the dynamic modulus of asphalt pavement [24]. By utilizing these methods and assessment criteria, the capacity of asphalt mixture to withstand permanent deformation can be more accurately determined. However, the complexity of the development of permanent deformation is due to many influential factors, such as binder type, asphalt content, mixture type, load level, temperature, etc. [10,17,20,24,25]. Therefore, it is crucial to choose the appropriate method to evaluate the permanent deformation in the design of asphalt mixtures.

According to the aforementioned studies, the asphalt mixtures' design is crucial in ensuring the performance of asphalt pavements. The purpose of this study is to choose a suitable design parameter for permanent deformation of bearing-layer asphalt mixture and establish a control standard. Intuitively, the layer's thickness can reflect the depth and distribution of the permanent deformation in each layer. We obtained core samples from the selected highway sections, calculated the contribution rate of permanent deformation for each layer, and identified the main layer and potential causes of the permanent deformation which had occurred in the asphalt pavement. An analysis was conducted using the local sensitivity method to evaluate the impact of asphalt type, gradation type, and asphalt content on the dynamic modulus, compressive strength, resilient modulus, shear strength, and splitting strength of bearing-layer asphalt mixtures. The significant performance parameter was chosen to establish a relationship with dynamic stability. Based on this, a control standard was suggested for the evaluation of the permanent deformation, one that connects the material design and the structural design of the bearing-layer asphalt mixture. This paper provides reasonable and effective guidance for prolonging the service life and improving the performance of asphalt pavement.

2. Investigation of Permanent Deformation of Asphalt Pavement in Northeast China

2.1. Road Section Information

The investigation of permanent deformation of asphalt pavement is the basis for analyzing the causes of the permanent deformation dysfunction and the structural function of the pavement. The Changyu Expressway in Jilin Province was selected (Figure 1a). It is one of the primary trunk highways, and has been in service for 12 years. The selected expressway is 143.5 km long and 26.0 m wide. A schematic diagram of the pavement's structure is shown in Figure 1b.

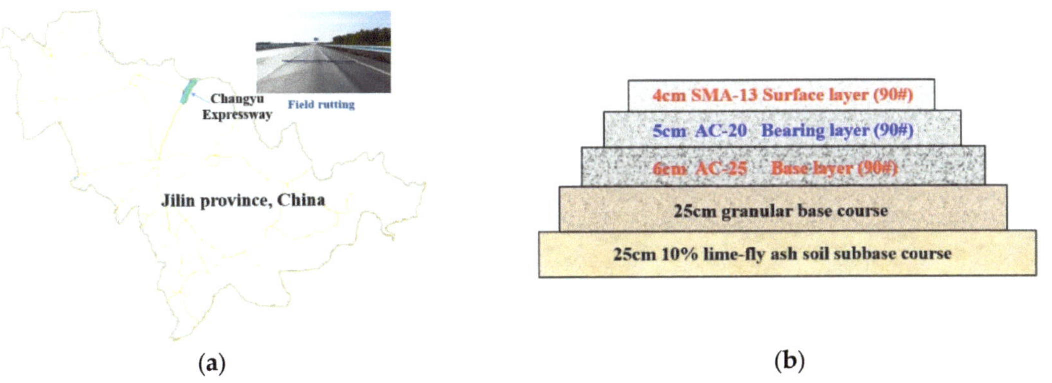

Figure 1. Information as to the selected expressway: (**a**) location and photograph of Changyu Expressway, (**b**) pavement structures.

The ZOYON-RTM intelligent road comprehensive detection vehicle (Figure 2a) was used to identify the permanent deformation state of Changyu Expressway. The detection system uses high-frequency, high-precision line-scan imaging 3D data acquisition technology to automatically obtain permanent deformation depth. The detection speed was 35 km/h and the temperature in the field was 30 °C. The permanent deformation depth in the typical pavement segment selected is shown in Figure 2b. The permanent deformation depth on the Changyu Expressway ranges from 6.31 mm to 26.75 mm, and the average value is 15.79 mm.

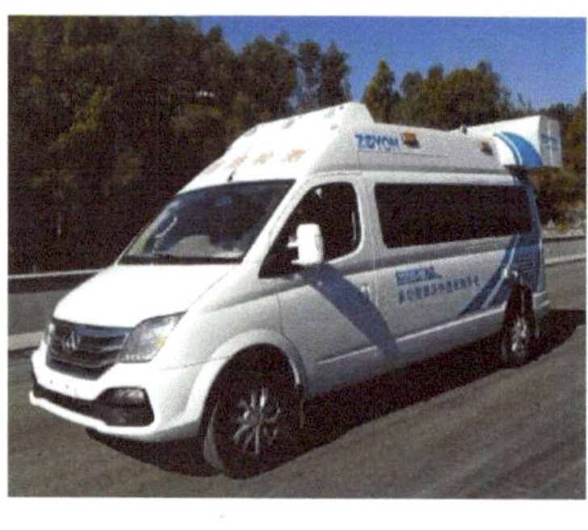

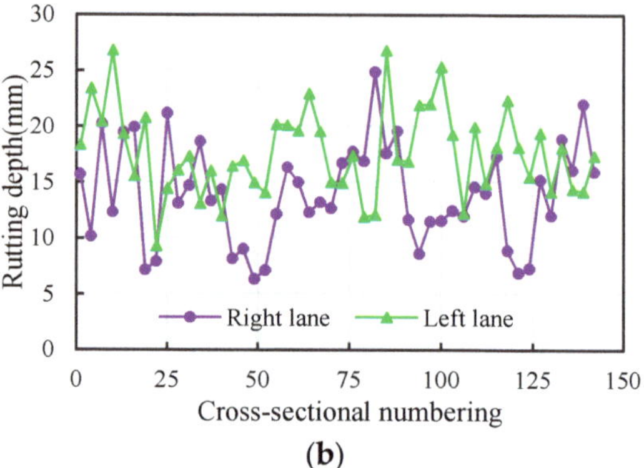

(a)   **(b)**

Figure 2. Field investigation: (**a**) ZOYON-RTM detection vehicle, (**b**) results of permanent deformation depth detection.

2.2. Field Cores Information

The aim was to enhance our understanding of the root causes of permanent deformation and to determine the contribution of each asphalt pavement layer to the permanent deformation. To achieve this goal, a set of full-depth core samples (with a diameter of 100 mm) were extracted. These samples were extracted from wheel paths in the same cross-section in the driving lane, the ridge band between wheel paths and shoulder, and the pavement shoulder, respectively. The layout scheme of field cores in a section is illustrated in Figure 3. One core (a) was drilled on a wheel path, one core (b) was drilled on the ridge band, and one core (c) was drilled on the road shoulder. Through examination of the changes in thickness within the field cores, the layer in the asphalt pavement where permanent deformation significantly occurs can be identified. Evidence from the literature shows that permanent deformation is closely related to the pavement materials and structures [11,15]. Hence, the design index for permanent deformation should be considered in the material design of the asphalt mixture located in that layer.

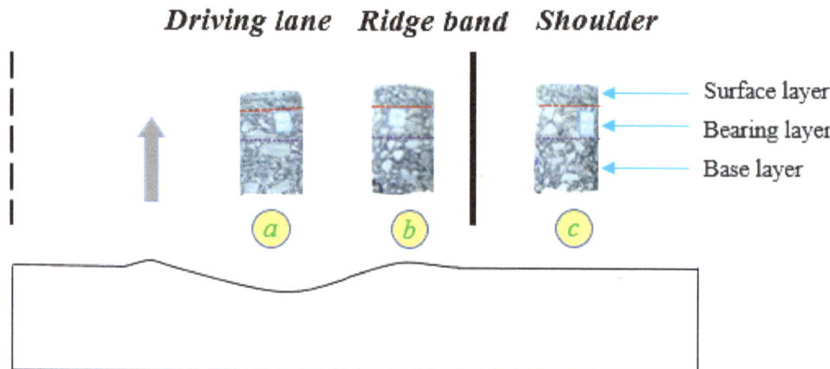

Figure 3. Schematic diagram for the selection of field cores.

3. Materials and Methods

3.1. Materials and Sample Preparation

The types of asphalt utilized in this study include 30# and 90# ordinary petroleum asphalt, as well as rubber-modified asphalt. The basic properties were tested following the Standard Test Methods of Bitumen and Bituminous Mixtures for Highway Engineering (JTG E20-2011) [26]. The results are presented in Table 1. The coarse and fine aggregates of the asphalt mixture were made of andesite, which is obtained from Heilongjiang Province. The mineral filler was crushed limestone. The properties of coarse aggregates and mineral filler are given in Tables 2 and 3, respectively. The technical indicators of the aggregate and mineral filler meet the requirements of the Technical Specification for Construction of Highway Asphalt Pavements (JTG/F40-2004) [27].

Table 1. Base asphalt properties.

Properties		Asphalt (30#)	Asphalt (90#)	Rubber-Modified Asphalt	Methods
Ductility (cm)	15 °C	56	>100	/	T0604-2011
	5 °C	/	/	41	
Penetration degree at 25 °C (0.1 mm)		29.0	83.7	65	T0605-2011
Softening point (°C)		73.2	51.4	65.6	T0606-2011
Dynamic viscosity at 60 °C (Pa·s)		742	187	13014	T0620-2011
Viscosity at 135 °C (Pa·s)		0.67	1.56	8.65	T0625-2011

Table 2. Coarse-aggregate properties.

Technical Indices	Results	Criteria	Methods
Crush value (%)	5	≤ 28	T0316-2005
Content of acicular and flaky shape particles (%)	8.6	≤ 15	T0312-2005
Losses of the Los Angeles abrasion test (%)	14.3	≤ 30	T0317-2005
Water absorption (%)	0.32	≤ 2	T0307-2005
Asphalt adhesion (graduation)	4	≥ 4	T0616-1993
Impact value (%)	17	≤ 30	T0322-2000
Firmness (%)	2.9	≤ 12	T0314-2000
Mud content (%)	0.8	≤ 1	T0310-2005

Table 3. Mineral-filler properties.

Properties	Hydrophilic Coefficient	Water Content (%)	Apparent Density (t/m³)	Size Distributions (%)		
				<0.075 mm	<0.15 mm	<0.6 mm
Results	0.634	0.5	2.720	80.7	96.3	100
Criteria	<1	≤ 1	≥ 2.50	75~100	90~100	100
Methods	T0353-2000	T0350-1994	T0352-2000		T0351-2000	

AC-20 mixture is commonly used in China for the construction of the bearing layer [14,16]. To clarify the influence of aggregate gradation on asphalt mixture performance, a comparative analysis was conducted on three AC-20 gradation options. The gradation composition of these asphalt mixtures is shown in Figure 4.

3.2. Test Methods

The purpose of choosing a property test index was to direct the asphalt mixture design towards resisting permanent deformation. As mentioned above, dynamic modulus is an important parameter used to measure the elastic properties of asphalt mixture. Uniaxial compressive strength can reflect the resistance of asphalt mixture to compressive deformation. Shear strength can represent the shear capacity of asphalt mixture. Splitting strength can measure the tensile capacity of asphalt mixture. Flow number and creep rate can reflect the ability of asphalt mixture to resist high-temperature deformation. The specific test methods are as follows.

3.2.1. Dynamic Modulus Test

A dynamic modulus test was carried out according to the Chinese specification JTG E20-2011. The specimens had heights of 150 mm and diameters of 100 mm. The test temperature was 25 °C. The frequencies were 25 Hz, 10 Hz, 5 Hz, 1 Hz, 0.5 Hz, and 0.1 Hz. The dynamic modulus has a certain stress dependence. When the dynamic loading frequency is constant, the dynamic modulus increases with an increase in stress [28]. Dynamic modulus was calculated as follows:

$$\left| E^* \right| = \frac{\sigma_0}{\varepsilon_0} \tag{1}$$

where $|E^*|$ is the dynamic modulus (MPa), σ_0 is the axial stress amplitude value (MPa), and ε_0 is the axial strain amplitude value (mm/mm).

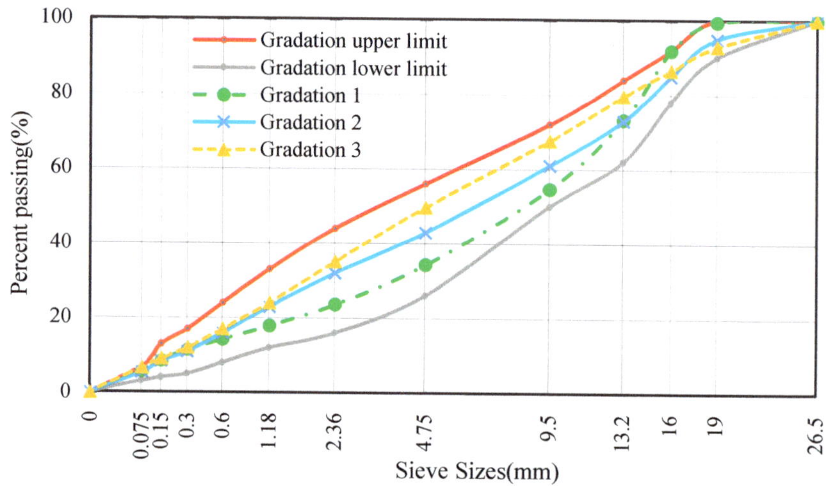

Figure 4. Aggregate gradation curves.

3.2.2. Uniaxial Compression Test

Uniaxial compression strength was obtained following the Chinese specification JTG E20-2011. The specimens had heights of 100 mm and diameters of 100 mm and had been manufactured on a shear gyratory compactor. The test temperature was 15 °C. The loading rate was 2 mm/min. There is a positive correlation between unconfined compressive strength and the capability to withstand permanent deformation. Uniaxial compression strength was obtained as follows:

$$R_c = \frac{P}{A} \tag{2}$$

where R_c is the uniaxial compression strength (MPa), P is the peak load at sample failure (N), and A is the cross-sectional area of the sample (mm^2).

3.2.3. Uniaxial Penetration Test

Shear strength was evaluated in accordance with the Specifications for Design of Highway Asphalt Pavement (JTG D50-2017) [29]. The specimens had heights of 100 mm and diameters of 100 mm and had been manufactured on a shear gyratory compactor. The test temperature was 60 °C. The loading rate was 1 mm/min. The greater the shear strength, the better the shear-deformation resistance of the asphalt mixture. Shear strength was computed as follows:

$$\tau_0 = \frac{fP}{A} \tag{3}$$

where τ_0 is the shear strength (MPa), f is the shear stress coefficient, P is the peak load at sample failure (N), and A is the cross-sectional area of the penetration mold (mm^2).

3.2.4. Splitting Test

Splitting strength was tested referring to the Chinese specification JTG E20-2011. The specimens were prepared by the standard Marshall method. The test temperature was 25 °C. The loading rate was 50 mm/min. The greater the splitting strength, the better the deformation resistance of the asphalt mixture. Splitting strength was calculated as follows:

$$R_T = \frac{0.006287 P_T}{h} \tag{4}$$

where R_T is the splitting strength (MPa), P_T is the failure load (N), and h is the specimen height (mm).

3.2.5. Dynamic Creep Test

The dynamic creep test is used to measure the flow number and creep rate of an asphalt mixture. To eliminate the effect on the Marshall sample during the initial loading stage, a preload of 0.002 MPa was applied for 10 min before loading. Then, a compressive stress of 0.05 MPa was applied. The loading waveform was a half sine wave, and the loading period was 1 s, including 0.1 s of loading and 0.9 s of unloading. The third stage of the test is characterized by the emergence of cracks in the specimen or a discernible deviation in the test curve, events which serve as the termination criteria [30]. Creep rate was obtained as follows:

$$\varepsilon_s = \frac{\varepsilon_2 - \varepsilon_1}{(t_2 - t_1)\sigma_0} \tag{5}$$

where ε_s is the creep rate (1/s/MPa), σ_0 is the flexural–tensile stress (MPa), and $t_1, t_2, \varepsilon_1, \varepsilon_2$ are the starting times and ending times of the stable period and the corresponding creep-strain values.

In this study, three types of asphalt (30# and 90# ordinary road petroleum asphalt, and rubber-modified asphalt), three AC-20 gradations (1, 2, and 3) and five asphalt content levels (3.9%, 4.2%, 4.5%, 4.8%, and 5.1%) were selected to investigate the influence on design parameters of the asphalt mixture. Three parallel sets of specimens were prepared for all tests that were carried out by the Universal Testing Machine-250 (IPC Global, Wantirna South, Australia), and the value of each test index was obtained by the average value of the three parallel specimens. Before testing, all the samples were stored in a temperature-controlled chamber at a specified temperature for 4~5 h.

4. Results and Discussion

4.1. Thickness Changes in Each Layer of the Field Cores

As mentioned in Section 2.2, in order to improve the ability of the asphalt pavement to resist permanent deformation, it is feasible to identify the layer where permanent deformation mainly occurs and optimize the design indices of this layer. Intuitively, the layer's thickness can reflect the depth and distribution of the permanent deformation in each layer. According to previous studies [10,11], the total pavement deformation in a typical pavement structure occurred solely in the three asphalt mixture layers above the granular base course. Hence, our analysis was limited to measuring the thicknesses of the three distinct layers to identify the layer where permanent deformation occurs most significantly. We selected five cross-sections with a permanent deformation depth of 15 mm and measured the thicknesses of the core specimens. The testing results for the thicknesses of each layer are shown in Figure 5.

From analysis of Figure 5, it can be determined that the reduction in thickness is more significant in the driving lane than in the ridge band and shoulder. This phenomenon can be attributed to the intensified densification and shearing deformation that arise from the combined impacts of recurrent axle loads and high temperatures [10,16]. The deformation of the asphalt mixture layer is the predominant cause of rutting. Specifically, for the driving lane, a minor alteration of 1.1 mm was observed in the thickness of the surface layer, which decreased from 40.0 to 38.9 mm. As a result of a thickness alteration in the bearing layer, the measurement decreased from 50.0 to 42.0 mm, which was accompanied by a variation of 8.0 mm. This change accounts for 53.3% of the total rutting depth. The thickness of the base layer changed from 60.0 to 55.1 mm, with a variation of 4.9 mm, accounting for approximately 32.7% of the whole rutting depth. In general, the bearing layer experiences the most severe rutting deformation and accounts for the majority of the deformation. Based on the available data, it appears that the bearing layer is the primary area of concern for rutting deformation in asphalt pavement. Therefore, it is imperative to incorporate the ability of bearing-layer asphalt mixture to withstand permanent deformation as a crucial design parameter.

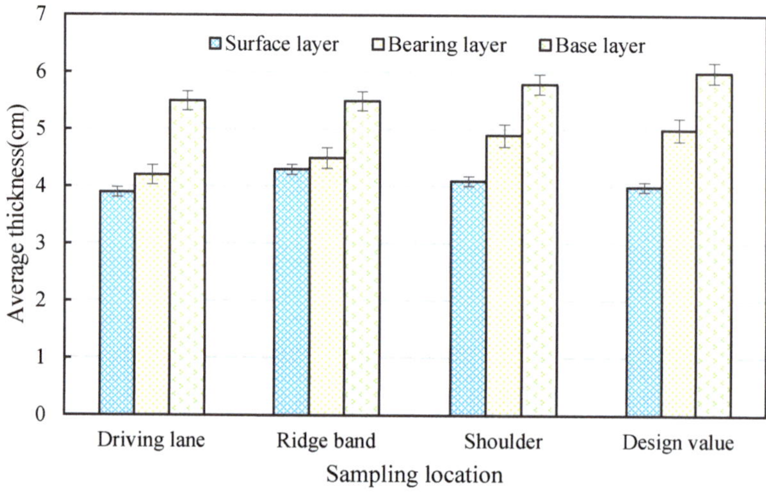

Figure 5. The thickness changes of each asphalt pavement layer.

4.2. Design Parameters of Asphalt Mixture in the Bearing Layer

At present, the relationship between asphalt mixture design and actual pavement performance is empirical. The Marshall method used in China is used for mixture design. Its design indices are volume parameters (porosity, asphalt saturation, mineral aggregate gap probability, etc.) and two mechanical indices (Marshall stability and flow value), which omits consideration of the structural role of the asphalt mixture in the bearing layer in the design index. This leads to a disconnection between the material design and the structural design of the bearing-layer asphalt mixture. In this study, we aim to obtain a mechanical index related to pavement performance in order to guide the material design of the bearing layer. This design index is determined by taking into account the dynamic modulus, compressive strength, resilient modulus, shear strength, and splitting strength.

4.2.1. Analysis of Factors Influencing Design Parameter

A comprehensive analysis of the mechanical indices of asphalt mixture was conducted when the gradation type, asphalt type, and asphalt contents were altered. Among them, the modulus of the material is an important parameter of the pavement structure [31]. Taking the dynamic modulus as an example, the results are shown in Figures 6 and 7.

Figure 6 presents the dynamic modulus curves at different frequencies, with an asphalt content of 4.5%. The dynamic modulus exhibits an increase as the frequency increases. The reason for this is that when the frequency is modified, the load will not be compressed to its full extent, and the "unload" will not bounce back entirely, resulting in energy buildup. As the frequency increases, the energy levels rise, and the strain decreases when subjected to constant stress. Therefore, the dynamic modulus will gradually increase.

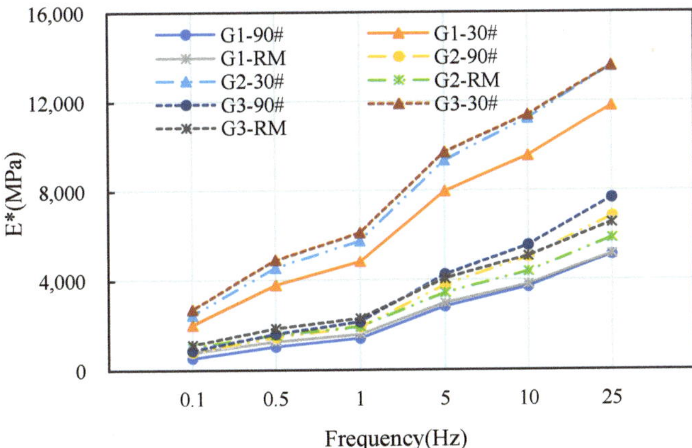

Figure 6. Dynamic modulus curves at different frequencies, with asphalt content of 4.5%. (Notes: G1-90# means asphalt mixture with gradation one and 90# asphalt; G means gradation; 30# means 30# asphalt; RM means rubber-modified asphalt).

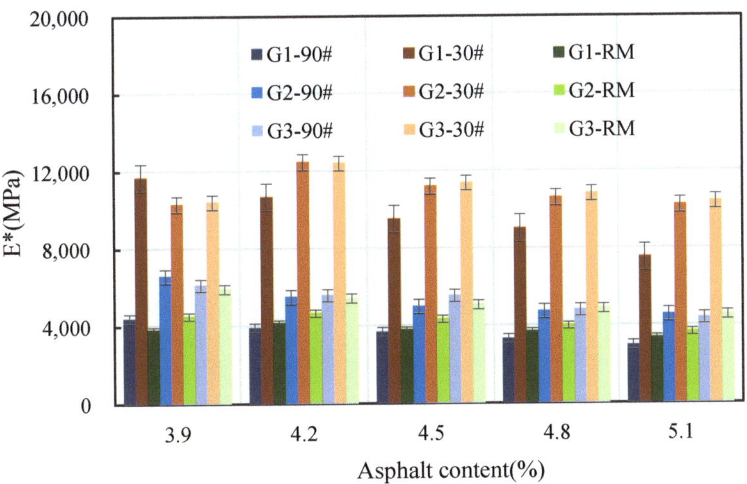

Figure 7. Dynamic modulus at different asphalt contents with frequency of 10 Hz.

Furthermore, the types of gradation and asphalt also have an impact on the dynamic modulus of asphalt mixtures. When comparing within gradation one, it was found that the dynamic modulus of 30# asphalt mixture was nearly double that of the 90# asphalt mixture with the same frequency and 4.5% asphalt content. However, the difference between the dynamic modulus of 90# asphalt mixture and rubber asphalt mixture was minimal. Similar results were obtained for asphalt mixtures when comparing within gradations two and three. For asphalt mixtures with the same type of asphalt but different gradations, the dynamic modulus has little difference under the condition of constant loading frequency and 4.5% asphalt content. Comparing the dynamic modulus values of gradations one, two, and three, it can be observed that gradation three has the highest modulus, followed by gradation two and then gradation one. It can be inferred that the dynamic modulus of asphalt mixtures is impacted by the type of gradation and asphalt.

A loading frequency of 10 Hz and a loading time of 0.016 s have been found to produce an effect on asphalt pavement comparable to that of a speed of 60–65 km/h. Therefore, the loading rate of 10 Hz is equivalent to the effects of the actual road speed [32]. To streamline the analysis process, the dynamic modulus at 10 Hz was adopted as the standard for analysis in subsequent tests. The dynamic modulus values at different asphalt contents with a frequency of 10 Hz are shown in Figure 7.

As shown in Figure 7, within a certain range of asphalt content, the dynamic modulus of 90# asphalt mixture shows a monotonically decreasing trend with an increase in asphalt content. For 30# asphalt and rubber-modified asphalt mixtures, the dynamic modulus initially increased and then decreased with the increase in asphalt content. And the inflection point occurs at 4.2% asphalt content. However, the dynamic modulus of the rubber-modified asphalt mixture with gradation three decreased monotonously with an increase in asphalt content. The amount of asphalt has the greatest impact on the dynamic modulus of the 90# asphalt mixture, while the dynamic modulus of the rubber asphalt mixture is the least affected by changes in the amount of asphalt. The dynamic modulus of the 30# asphalt mixture falls between these two extremes. This shows that the content of asphalt has a great influence on the dynamic modulus of asphalt mixtures.

The test results of other design indices are shown in Table 4, including compressive strength, resilient modulus, shear strength, and splitting strength. The index value of each experimental condition was obtained from the mean value of three parallel experiments. It can be seen that these indices are significantly affected by asphalt type and gradation type. For asphalt type, under the condition of a certain gradation and asphalt content, the compressive strength, resilient modulus, and splitting strength of 30# asphalt mixture are the highest, followed by those of rubber-modified asphalt mixture, and 90# asphalt mixture returns the lowest values. The shear strength of rubber-modified asphalt mixture is the highest, followed by 30# asphalt mixture, and 90# asphalt mixture returns the lowest values. Under the condition of a certain asphalt type and asphalt content, comparing these indices of gradations one, two, and three, it can be observed that gradation three has the highest values, followed by gradation two and then gradation one. As to the content level of asphalt, with an increase in asphalt content under the set condition of a certain gradation and asphalt type, these mechanical indices initially increase and then decrease. The peak points are mainly at 4.2% and 4.5% asphalt content, findings which are basically consistent with the conclusion of the dynamic modulus. This also indicates that the optimal amounts of asphalt for asphalt mixtures with different gradations and types are 4.2% and 4.5%. The above analysis shows that gradation, asphalt type, and asphalt content have effects on the mechanical index of an asphalt mixture.

Table 4. List of design index test results.

Gradation Type	Asphalt Content (%)	Compressive Strength (MPa)			Resilient Modulus (MPa)			Shear Strength (MPa)			Splitting Strength (MPa)		
		90#	30#	RM	90#	30#	RM	90#	30#	RM	90#	30#	RM
Gradation 1	3.9	3.08	6.54	3.74	1293	2581	1173	1.07	1.55	1.52	1.01	1.63	0.87
	4.2	3.23	6.80	3.91	1391	2752	1315	0.99	1.48	1.50	1.19	1.67	0.86
	4.5	2.96	6.73	4.11	1388	2609	1243	0.96	1.33	1.63	1.14	1.71	0.93
	4.8	2.86	6.71	4.39	1235	2407	1162	0.89	1.31	1.45	1.00	1.63	0.96
	5.1	2.75	6.45	3.82	1094	2378	1091	0.85	1.22	1.38	0.94	1.54	0.95
Gradation 2	3.9	2.78	5.89	3.37	1165	2326	1056	1.73	2.46	2.62	0.69	1.64	0.96
	4.2	3.38	7.12	4.09	1456	2880	1376	1.81	2.51	2.66	0.71	1.76	1.04
	4.5	3.02	6.87	4.20	1419	2666	1270	1.98	2.92	2.79	0.79	1.83	1.08
	4.8	3.23	7.59	4.96	1397	2721	1313	1.88	2.67	2.89	0.75	1.77	1.00
	5.1	3.12	7.32	4.33	1243	2701	1238	1.81	2.57	3.07	0.66	1.58	0.97
Gradation 3	3.9	3.58	5.33	3.51	1323	2138	1110	1.67	1.89	2.49	0.89	1.63	1.03
	4.2	3.89	5.86	3.99	1646	2367	1234	2.11	2.06	2.56	1.00	1.74	1.05
	4.5	3.38	5.87	3.78	1528	2104	1326	2.14	2.15	2.61	1.09	1.72	1.10
	4.8	3.26	5.52	3.83	1460	1966	1270	1.96	2.06	2.80	0.98	1.64	1.23
	5.1	3.10	5.07	3.62	1256	2021	1093	1.83	1.94	2.56	0.94	1.57	1.07

4.2.2. Sensitivity Analysis

The sensitivity of each design index to asphalt type, gradation type, and asphalt content was analyzed by the single-factor analysis of variance method. The calculation results are shown in Table 5. The results of the single-factor analysis of variance were subjected to a 95% confidence level. A variable can be deemed to have a significant impact on the design index value of the asphalt mixture if its significance index p value is less than 0.05 [19].

Table 5. The sensitivity of design indices to asphalt type, gradation type, and asphalt content.

Gradation Type	Asphalt Type	Dynamic Modulus		Compressive Strength		Resilient Modulus		Shear Strength		Splitting Strength	
		p	Significance	p	Significance	p	Significance	p	Significance	p	Significance
Gradation 1	90#	0.0041	**	0.0523	-	0.1710	-	0.1771	-	0.0204	*
	30#	0.0008	***	0.4557	-	0.0588	-	0.0689	-	0.4392	-
	RM	0.0483	*	0.0492	*	0.1585	-	0.3581	-	0.2447	-
Gradation 2	90#	0.0057	**	0.1127	-	0.1591	-	0.2366	-	0.1109	-
	30#	0.0206	*	0.0733	-	0.1675	-	0.0648	-	0.1105	-
	RM	0.0483	*	0.1516	-	0.0443	*	0.0450	*	0.3659	-
Gradation 3	90#	0.0210	*	0.0325	*	0.1522	-	0.0427	*	0.2240	-
	30#	0.0268	*	0.0991	-	0.0579	-	0.0819	-	0.4635	-
	RM	0.0518	-	0.3828	-	0.3159	-	0.1655	-	0.2329	-

Notes: *** means highly significant; ** means mid-level significance; * means low significance; - means not significant.

According to the results of the single-factor variance analysis in Table 5, the dynamic modulus is sensitive to a change of asphalt content within the same gradation and asphalt type. The splitting strength is not sensitive to changes in asphalt content. The compressive strength, resilient modulus, and shear strength are partially sensitive to changes in asphalt content, but partially insensitive. Therefore, it is reasonable to take the dynamic modulus as the design index for asphalt mixtures. Meanwhile, in the case of a limited sample size of asphalt mixture variations, the type of asphalt had the most significant influence on the dynamic modulus. This means that the type of asphalt has an important effect on the permanent deformation of asphalt mixtures. Compared with modifying the gradation type or the asphalt content, changing the type of asphalt is more helpful in improving the resistance to the permanent deformation of the asphalt mixture.

4.3. Control Standard for Asphalt Mixture in the Bearing Layer

The connection between permanent deformation and dynamic modulus has been poorly investigated in existing studies. Through extensive research on the correlation between dynamic creep and dynamic modulus, the connections between permanent deformation, dynamic modulus, and dynamic creep have been established by these findings. As a result, the dynamic modulus is proposed as the design index for bearing-layer asphalt mixture in accordance with different permanent-deformation control standards.

4.3.1. Analysis of Factors Influencing Dynamic Creep

Empirical evidence suggests that the dynamic creep test has a very strong correlation with permanent deformation depth and a high capability to estimate the potential of permanent deformation [33]. The dynamic creep test has been used to judge and predict permanent deformation, and to investigate the creep law and extent of deformation of asphalt mixtures under repeated loads [34,35]. According to existing research results, creep development of asphalt mixtures has three distinct strain stages [36]: (1) primary stage, in which the strain rate decreases; (2) secondary stage, in which the strain rate is constant; and (3) tertiary stage, in which the strain rate increases. The loading cycle corresponding to the deformation curve upon reaching the third stage is defined as the flow number (F_N). During the stable period, the rate at which strain increases per unit of time under a unit of stress is known as the creep rate. The flow number and creep rate are commonly used as indicators to evaluate the stability of asphalt mixtures at high temperatures [37]. Figure 8 shows an example of a creep curve.

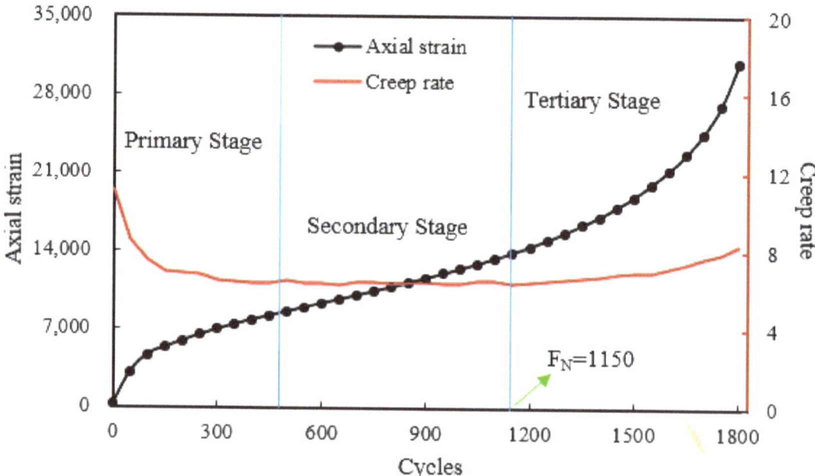

Figure 8. Three stages of creep development.

In this study, the impact of dynamic creep was examined using three types of asphalt (30# and 90# ordinary petroleum asphalt, and rubber-modified asphalt), five asphalt content levels (3.9%, 4.2%, 4.5%, 4.8%, and 5.1%), and three AC-20 gradations (one, two, and three). The flow number and creep rate at different asphalt contents are shown in Figures 9 and 10, respectively.

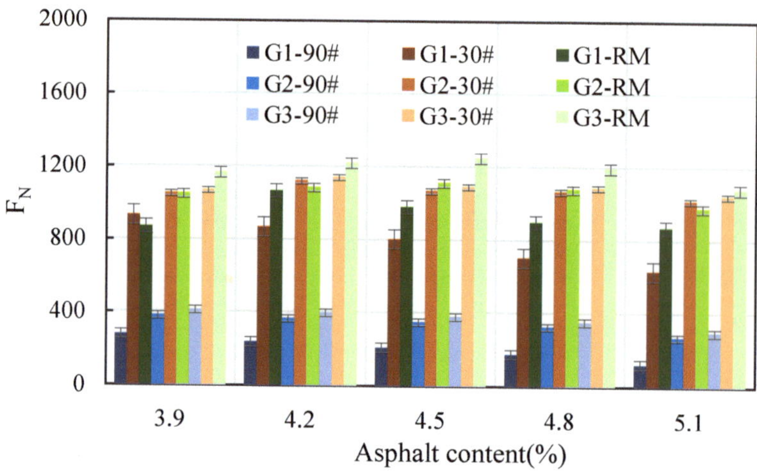

Figure 9. Flow number at different asphalt contents.

The flow number is used to describe the permanent-deformation behavior of an asphalt mixture under loading stress with time. A higher flow number indicates a greater resistance to permanent deformation for the asphalt mixture. As seen in Figure 9, for asphalt mixtures with the same gradation and asphalt content, the flow number of the 30# ordinary asphalt and rubber-modified asphalt mixture were not significantly different, but they were significantly higher than that of the 90# asphalt mixture. This indicates that

90# asphalt mixture has poor high-temperature resistance. For the asphalt mixture with the same asphalt content, the flow number of gradation 1 was the smallest, followed by that of gradation two, and the flow number of gradation three was the largest. Within a certain range of asphalt content, the flow number of the 90# asphalt mixture decreased monotonically with an increase in asphalt content. For the 30# asphalt, both alone and in rubber-modified asphalt mixtures, the flow number of the 30# asphalt mixture with gradation one decreased monotonically with an increase in asphalt content. However, the flow numbers of the other mixtures first increased and then decreased with an increase in asphalt content. The inflection points appeared at the asphalt contents of 4.2% and 4.5%. The results indicate that the flow number of asphalt mixture is significantly affected by the asphalt type, gradation type, and asphalt content.

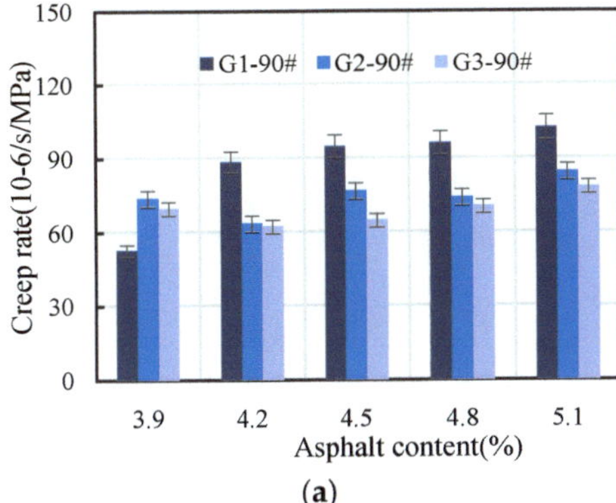

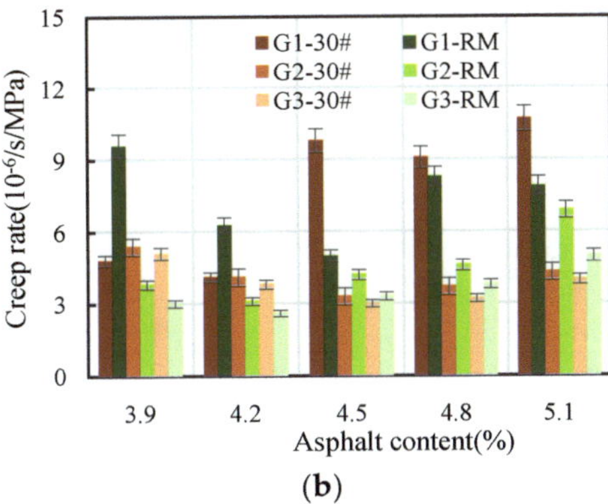

Figure 10. Creep rate at different asphalt contents: (**a**) 90#, (**b**) 30# and RM.

Creep rate serves as an indicator of the permanent-deformation performance of asphalt mixture under loading stress with time. The smaller the creep rate, the stronger the permanent-deformation resistance of the asphalt mixture. As shown in Figure 10, for asphalt mixtures with the same gradation and asphalt content, the creep rate of 30# ordinary asphalt and rubber-modified asphalt mixture showed little difference, but they were significantly lower than that of the 90# asphalt mixture. This also indicates that the permanent deformation resistance of 90# asphalt mixture is poor. For asphalt mixtures of the same type of asphalt and asphalt content level, the general rule was that the creep rate of gradation one was the highest, followed by that of gradation two, and the creep rate of gradation three was the lowest. Within a certain range of asphalt content, the creep rate of asphalt mixture varied depending on the type of asphalt. Except for the 90# asphalt mixture with gradation one, whose creep rate increased monotonically with an increase in asphalt content, the creep rate of other asphalt mixtures first decreased and then increased with the increase in asphalt content. The inflection point occurred when the asphalt content was between 4.2% and 4.5%. The results show that the creep rate of asphalt mixture is also greatly affected by asphalt type, gradation type, and asphalt content.

4.3.2. Relationship between Dynamic Modulus and Permanent Deformation

As evident from the outcomes mentioned earlier, the creep rate and flow number exhibit clear trends with variations in asphalt content, and effectively differentiate the permanent deformation of asphalt mixture. Based on the experimental results for dynamic modulus, the correlation between creep parameters and dynamic modulus indices is analyzed below. Among these values, the dynamic modulus at 10 Hz is selected, because the loading rate of 10 Hz is equivalent to the effects of the actual road speed [32]. The relationship between the dynamic modulus and creep parameters of ordinary petroleum asphalt mixtures and rubber-modified asphalt mixtures, respectively, is shown in Figures 11 and 12.

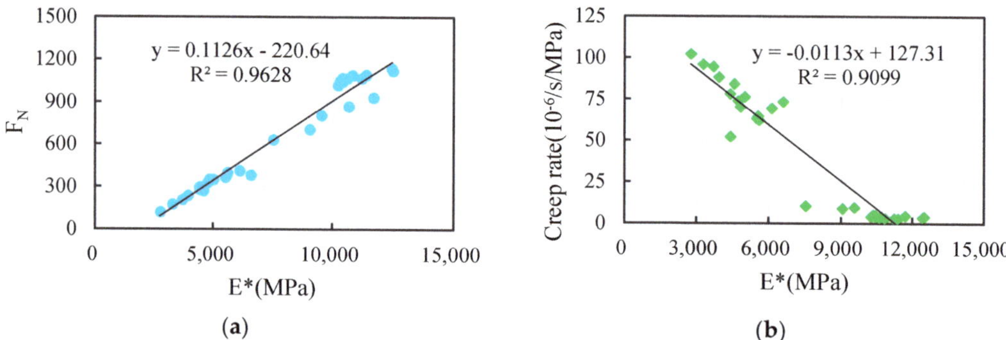

Figure 11. Relationship between dynamic modulus and creep parameters with 30# and 90#: (**a**) flow number, (**b**) creep rate.

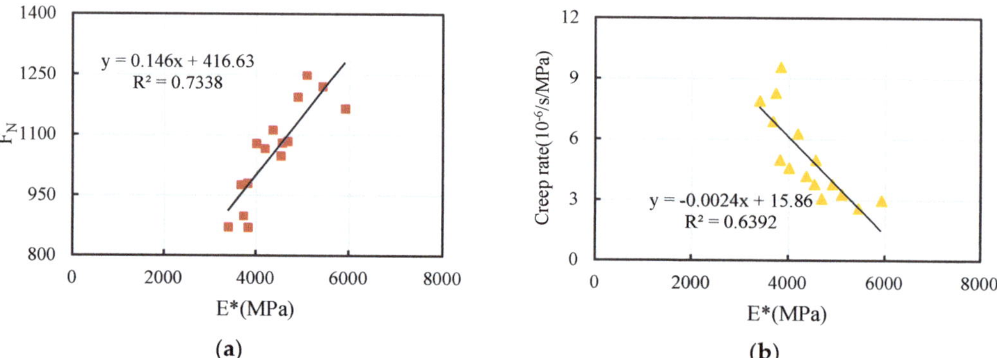

Figure 12. Relationship between dynamic modulus and creep parameters with RM: (**a**) flow number, (**b**) creep rate.

Figure 11a conveys that the flow number of ordinary petroleum asphalt mixture has a strong correlation with the dynamic modulus. The flow number increases linearly with the increase in dynamic modulus. The correlation coefficient between them is 0.9628. As shown in Figure 11b, it is evident that the creep rate exhibited by ordinary petroleum asphalt mixture is positively correlated with the dynamic modulus. The creep rate decreases linearly with an increase in dynamic modulus. The correlation coefficient between them is 0.9099.

It can be seen from Figure 12a that the flow number of rubber-modified asphalt mixture has a certain correlation with the dynamic modulus. The flow number increases with an increase in dynamic modulus. The correlation coefficient between them is 0.7338. The correlation between the creep rate of rubber-modified asphalt mixture and the dynamic modulus is evident in Figure 12b. The creep rate decreases with an increase in dynamic modulus. The correlation coefficient between them is 0.6392.

As evidenced by Figures 11 and 12, the relationship between creep parameters and dynamic modulus is more pronounced for ordinary asphalt mixture in comparison to rubber-modified asphalt mixture. The correlation between flow number and dynamic modulus is higher than that between creep rate and dynamic modulus. Therefore, the flow number is chosen as the assessment criterion for creep testing in order to carry out standardized investigations on dynamic modulus parameters.

In the literature, the analysis of dynamic creep test results and rutting test results performed by Qi Feng indicates that there is a good relationship between the flow number and permanent deformation [38]. The relationship is as follows (the correlation coefficient R^2 is 0.927):

$$DS = 2.8099 \cdot F_N + 535.6 \tag{6}$$

According to the analysis in Figure 12a, when the test temperature is 25 °C and the loading frequency is 10 Hz, the relationship between the flow number and the dynamic modulus is as follows (the correlation coefficient R^2 is 0.963):

$$F_N = 0.1126 \cdot E^* - 220.6 \tag{7}$$

The connection between permanent deformation and dynamic modulus can be derived by merging Formulas (6) and (7).

$$DS = 0.3146 \cdot E^* - 84.3 \tag{8}$$

By utilizing the flow numbers and dynamic modulus test results at different frequencies determined at 25 °C, the correlation between permanent deformation and dynamic modulus can be determined. The results are summarized in Table 6.

Table 6. Dynamic modulus values under different control standards. (MPa).

Frequency (Hz)	Permanent Deformation Control Standard (Times/mm)						Relational Expression
	800	2400	2800	3000	4000	5000	
25	4462	10,207	11,643	12,362	15,952	19,543	$DS = 0.2785 \times |E^*| - 442$
10	2794	7851	9115	9747	12,908	16,068	$DS = 0.3164 \times |E^*| - 84$
5	2027	6616	7763	8336	11,204	14,072	$DS = 0.3487 \times |E^*| + 93$
1	682	3890	4692	5093	7098	9103	$DS = 0.4988 \times |E^*| + 460$
0.5	424	3041	3695	4022	5658	7293	$DS = 0.6114 \times |E^*| + 541$
0.1	163	1624	1989	2172	3085	3998	$DS = 1.0953 \times |E^*| + 621$

The results presented in Table 6 demonstrate that when frequency and temperature remain constant, an increase in dynamic stability leads to a higher dynamic modulus. By referring to Table 6 or utilizing an interpolation method, one can acquire the dynamic modulus values of asphalt mixtures according to various permanent-deformation control standards. For example, when the permanent deformation of asphalt mixture is required to be $\geq$2400 times/mm, it can be found by using Table 6 that $E^* \geq 7851$ MPa is obtained when the loading frequency is 10 Hz. This is basically consistent with the requirements of the Chinese specification JTG D50-2017.

5. Conclusions

This study focused on an investigation of the relationship between permanent deformation and dynamic modulus performance in bearing-layer asphalt mixtures. The following conclusions were drawn.

Intuitively, the layer's thickness can reflect the depth and distribution of the permanent deformation in each layer. Through field investigation and coring, it was determined that the reduction in thickness is more significant in the driving lane than in the ridge band and shoulder. This phenomenon can be attributed to the intensified densification and shearing deformation that arise from the combined impacts of recurrent axle loads and high temperatures. Compared to surface and base layers, the bearing layer is the primary area of concern for rutting deformation in asphalt pavement. Therefore, it is necessary to strengthen the evaluation of the high-temperature performance of bearing-layer asphalt mixture.

By sensitively analyzing the results, it can be determined that the dynamic modulus of the bearing-layer asphalt mixture is significantly influenced by the type of asphalt, gradation type, and asphalt content, compared to other design parameters. The relationships between dynamic modulus, flow number, and permanent deformation were combined to establish the control standard for a bearing-layer asphalt mixture. The dynamic modulus should not be less than 7851 MPa when the permanent deformation of asphalt mixture is required to be ≥2400 times/mm. This research can provide a mechanical index associated with pavement performance, one which can be utilized to guide material design and consequently extend the lifespan of the pavement, as well as enhance its performance.

Author Contributions: W.J.: investigation, methodology, formal analysis, writing—original draft; Y.M.: data curation, validation; Y.S.: investigation, data curation; X.Z.: conceptualization, data curation, validation; H.X.: validation, supervision, writing—review and editing. All authors have read and agreed to the published version of the manuscript.

Funding: This paper was financed by the National Key Research and Development Program of China (Grant No. 2022YFB2602600), the National Natural Science Funds of China (Grant No. 52278449), Heilongjiang Natural Science Funds of China (ZD2003E003), and Ningxia Natural Science Funds of China (2022AAC03759).

Institutional Review Board Statement: Not applicable.

Informed Consent Statement: Not applicable.

Data Availability Statement: The data presented in this study are available on request from the corresponding author.

Acknowledgments: Thanks to the anonymous reviewers for their comments that have notably helped us improve the manuscript.

Conflicts of Interest: The authors declare no conflict of interest.

References

1. Liu, G.; Chen, L.; Qian, Z.; Zhang, Y.; Ren, H. Rutting Prediction Models for Asphalt Pavements with Different Base Types Based on Riohtrack Full-Scale Track. *Constr. Build. Mater.* **2021**, *305*, 124793. [CrossRef]
2. Li, H.; Xu, H.; Chen, F.; Liu, K.; Tan, Y.; Leng, B. Evolution of water migration in porous asphalt due to clogging. *J. Clean. Prod.* **2022**, *330*, 129823. [CrossRef]
3. Liu, Y.; Su, P.; Li, M.; You, Z.; Zhao, M. Review on evolution and evaluation of asphalt pavement structures and materials. *J. Traffic Transp. Eng. (Engl. Ed.)* **2020**, *7*, 573–599. [CrossRef]
4. Garbowski, T.; Pożarycki, A. Multi-level backcalculation algorithm for robust determination of pavement layers parameters. *Inverse Probl. Sci. Eng.* **2017**, *25*, 674–693. [CrossRef]
5. Mataei, B.; Nejad, F.M.; Zahedi, M.; Zakeri, H. Evaluation of pavement surface drainage using an automated image acquisition and processing system. *Automat. Constr.* **2018**, *86*, 240–255. [CrossRef]
6. Wang, S.; Sui, X.; Leng, Z.; Jiang, J.; Lu, G. Asphalt pavement density measurement using non-destructive testing methods: Current practices, challenges, and future vision. *Constr. Build. Mater.* **2022**, *344*, 128154. [CrossRef]
7. Garbowski, T.; Gajewski, T. Semi-automatic inspection tool of pavement condition from three-dimensional profile scans. *Procedia Eng.* **2017**, *172*, 310–318. [CrossRef]
8. De Blasiis, M.R.; Di Benedetto, A.; Fiani, M. Mobile laser scanning data for the evaluation of pavement surface distress. *Remote Sens.* **2020**, *12*, 942. [CrossRef]
9. Yao, L.; Leng, Z.; Jiang, J.; Ni, F.; Zhao, Z. Nondestructive Prediction of Rutting Resistance of in-Service Middle Asphalt Layer Based on Gene Expression Programing. *Constr. Build. Mater.* **2021**, *293*, 123481. [CrossRef]
10. Xu, T.; Huang, X. Investigation into Causes of in-Place Rutting in Asphalt Pavement. *Constr. Build. Mater.* **2012**, *28*, 525–530. [CrossRef]
11. Li, N.; Zhan, H.; Yu, X.; Tang, W.; Yu, H.; Dong, F. Research on the High Temperature Performance of Asphalt Pavement Based on Field Cores with Different Rutting Development Levels. *Mater. Struct.* **2021**, *54*, 70. [CrossRef]
12. Azarhoosh, A.; Koohmishi, M. Investigation of the Rutting Potential of Asphalt Binder and Mixture Modified by Styrene-Ethylene/Propylene-Styrene Nanocomposite. *Constr. Build. Mater.* **2020**, *255*, 119363. [CrossRef]
13. Imaninasab, R.; Loria-Salazar, L.; Carter, A. Impact of Aggregate Structure Restoration on Rutting Resistance of Asphalt Mixtures with Very High Percentages of Rap. *Road Mater. Pavement Des.* **2023**, 1–17, *ahead-of-print*. [CrossRef]
14. Li, S.; Fan, M.; Xu, L.; Tian, W.; Yu, H.; Xu, K. Rutting Performance of Semi-Rigid Base Pavement in Riohtrack and Laboratory Evaluation. *Front. Mater.* **2021**, *7*, 590604. [CrossRef]
15. Zhao, Z.; Jiang, J.; Ni, F.; Dong, Q.; Ding, J.; Ma, X. Factors Affecting the Rutting Resistance of Asphalt Pavement Based on the Field Cores Using Multi-Sequenced Repeated Loading Test. *Constr. Build. Mater.* **2020**, *253*, 118902. [CrossRef]
16. Zhu, H.; Cai, H.; Ren, Q.; Hong, S.; Cao, R. Experiment and Analysis on the Development Law of High Temperature Performance of Existing Asphalt Pavement. *Highway* **2020**, *65*, 17–22.

17. Baghaee Moghaddam, T.; Baaj, H. The Use of Compressible Packing Model and Modified Asphalt Binders in High-Modulus Asphalt Mix Design. *Road Mater. Pavement Des.* **2020**, *21*, 1061–1077. [CrossRef]
18. Guo, W.; Guo, X.; Chang, M.; Dai, W. Evaluating the Effect of Hydrophobic Nanosilica on the Viscoelasticity Property of Asphalt and Asphalt Mixture. *Materials* **2018**, *11*, 2328. [CrossRef]
19. Ma, X.; Zhou, P.; Jiang, J.; Hu, X. High-Temperature Failure of Porous Asphalt Mixture Under Wheel Loading Based On 2D Air Void Structure Analysis. *Constr. Build. Mater.* **2020**, *252*, 119051. [CrossRef]
20. Zhang, Q.; Wang, S.; Lv, R.; Wu, J.; Qi, H. Viscoelastic Mechanical Performance of Dense Polyurethane Mixtures Based on Dynamic and Static Modulus Testing and Creep Testing. *Constr. Build. Mater.* **2022**, *320*, 126207. [CrossRef]
21. NCHRP. *Simple Performance Test for Superpave Mix Design*; (Project 9-19 Fy'98); NCHRP REPORT 465; Transportation Research Board: Washington, DC, USA, 2002.
22. Zhang, J.; Alvarez, A.E.; Lee, S.I.; Torres, A.; Walubita, L.F. Comparison of Flow Number, Dynamic Modulus, and Repeated Load Tests for Evaluation of HMA Permanent Deformation. *Constr. Build. Mater.* **2013**, *44*, 391–398. [CrossRef]
23. Chaturabong, P.; Bahia, H.U. Mechanisms of Asphalt Mixture Rutting in the Dry Hamburg Wheel Tracking Test and the Potential to be Alternative Test in Measuring Rutting Resistance. *Constr. Build. Mater.* **2017**, *146*, 175–182. [CrossRef]
24. Witczak, M.W.; Kaloush, K.E.; Von Quintus, H. Pursuit of the Simple Performance Test for Asphalt Mixture Rutting. *J. Assoc. Asph. Paving Technol.* **2002**, *71*, 671–691.
25. Xu, H.; Zhan, H.; Wang, M.; Tan, Y. Microstructural characteristics evolution of asphalt mastics under cyclic environmental temperature variations. *Constr. Build. Mater.* **2023**, *400*, 132795. [CrossRef]
26. JTG E20-2011; Standard Test Methods of Bitumen and Bituminous Mixtures for Highway Engineering. MOT: Beijing, China, 2011.
27. JTG/F40-2004; Technical Specification for Construction of Highway Asphalt Pavements. MOT: Beijing, China, 2011.
28. Xu, H.; Fu, X.; Meng, A.; Cui, H.; Tan, Y. Rayleigh wave technology for analysis dynamic modulus of asphalt mixtures: Feasibility and application. *Int. J. Pavement Eng.* **2020**, *21*, 1235–1247. [CrossRef]
29. JTG D50-2017; Specifications for Design of Highway Asphalt Pavement. MOT: Beijing, China, 2017.
30. Ma, Y.; Wang, H.; Zhao, K.; Yan, L.; Yang, D. Study of a Modified Time Hardening Model for the Creep Consolidation Effect of Asphalt Mixtures. *Materials* **2022**, *15*, 2710. [CrossRef]
31. Harnaeni, S.R.; Pramesti, F.P.; Budiarto, A.; Setyawan, A.; Khan, M.I.; Sutanto, M.H. Study on Structural Performance of Asphalt Concrete and Hot Rolled Sheet through Viscoelastic Characterization. *Materials* **2020**, *13*, 1133. [CrossRef]
32. Xu, Z.; Chang, Y.; Chen, Z.; Sun, J.; Liu, Y. Study on Test Standard of Asphalt Mixture Dynamic Modulus. *J. Traffic Transp. Eng.* **2015**, *15*, 1–8.
33. Kaloush, K.E.; Witczak, M.W. Tertiary Flow Characteristics of Asphalt Mixtures. *J. Assoc. Asph. Paving Technol.* **2002**, *71*, 278–306.
34. Gu, X.; Zhang, X.; Lv, J. Establishment and Verification of Prediction Models of Creep Instability Points of Asphalt Mixtures at High Temperatures. *Constr. Build. Mater.* **2018**, *171*, 303–311. [CrossRef]
35. Du, Y.; Dai, M.; Deng, H.; Deng, D.; Wei, T.; Kong, L. Laboratory Investigation on Thermal and Road Performances of Asphalt Mixture Containing Glass Microsphere. *Constr. Build. Mater.* **2020**, *264*, 120710. [CrossRef]
36. Khodaii, A.; Mehrara, A. Evaluation of Permanent Deformation of Unmodified and SBS Modified Asphalt Mixtures Using Dynamic Creep Test. *Constr. Build. Mater.* **2009**, *23*, 2586–2592. [CrossRef]
37. Büchner, J.; Wistuba, M.P. Assessing Creep Properties of Asphalt Binder, Asphalt Mastic and Asphalt Mixture. *Road Mater. Pavement Des.* **2022**, *23*, 116–130. [CrossRef]
38. Qi, F. Evaluation of High Temperature Stability of Asphalt Mixture by Creep Test. Master's Thesis, Chang'an University, Xi'an, China, 2009.

MDPI AG
Grosspeteranlage 5
4052 Basel
Switzerland
Tel.: +41 61 683 77 34

Materials Editorial Office
E-mail: materials@mdpi.com
www.mdpi.com/journal/materials

www.ingramcontent.com/pod-product-compliance
Lightning Source LLC
LaVergne TN
LVHW071939160726
843515LV00011B/2649